Art and Biotechnology

Viral Culture from CRISPR to COVID

Biotechne: Interthinking Art, Science and Design

Series Editors

Charissa N. Terranova (University of Texas at Dallas, USA)

Meredith Tromble (San Francisco Art Institute, USA)

Biotechne: Interthinking Art, Science and Design publishes books about the history, theory and practice of art and design as they comingle with the natural sciences. The series title reclaims the Greek meanings of the roots bios, conveying life, the living, or citizen-life, and techne, conveying art, skill, or craft. Biotechne thus names the folding of "art" and "science" into complex and hybrid practices that transcend a human-centered "engineering" worldview. "Interthinking," a neologism invented by art and science visionary György Kepes, describes knowledge informed by ecological, systemic, and cybernetic connections, defining the active engagement among fields central to the Biotechne series. This engagement is the source of the cultural creativity and resourcefulness necessary to thrive in the rapidly changing world conditions of the Anthropocene. Biotechne welcomes proposals treating art and design subjects from any time period, antiquity to the present, which speak directly to these contemporary concerns. We seek inventive, cross-pollinating works about the arts and their engagement with sciences from astrobiology to zoology, wherever that engagement occurs, in art or design studios, scientific laboratories, natural habitats, the museum and gallery worlds, performance spaces, medical practices, and the political realm. By identifying significant intersections of art, humanities, and science, and tracking rigorous paths through the cross-disciplinary information jungle, Biotechne serves audiences of both experts and lay readers while substantiating the role of aesthetic insight within the natural sciences.

Titles in the Series

D'Arcy Wentworth Thompson's Generative Influences in Art, Design, and Architecture: From Forces to Forms, edited by Ellen K. Levy and Charissa N. Terranova (2021)

Sea Currents in Nineteenth-Century Art, Science and Culture: Commodifying the Ocean World, edited by Molly Duggins and Kathleen Davidson (2023)

Plants by Numbers: Art, Computation, and Queer Feminist Technoscience, edited by Jane Prophet and Helen V. Pritchard (2023)

Space Feminisms: People, Planets, Power, edited by Marie-Pier Boucher, Claire Webb, Annick Bureaud, and Nahum (2024)

Art and Biotechnology

Viral Culture from CRISPR to COVID

Edited by Claire Correo Nettleton and Louise Mackenzie

BLOOMSBURY VISUAL ARTS
LONDON • NEW YORK • OXFORD • NEW DELHI • SYDNEY

BLOOMSBURY VISUAL ARTS
Bloomsbury Publishing Plc
50 Bedford Square, London, WC1B 3DP, UK
1385 Broadway, New York, NY 10018, USA
29 Earlsfort Terrace, Dublin 2, Ireland

BLOOMSBURY, BLOOMSBURY VISUAL ARTS and the Diana logo are trademarks of Bloomsbury Publishing Plc

Cover design: Elena Durey
Cover image: Stephanie Rothenberg, *Aphrodisiac in the Machine*, detail © Stephanie Rothenberg

A catalogue record for this book is available from the British Library.

Library of Congress Cataloging-in-Publication Data
Names: Nettleton, Claire, editor. | Mackenzie, Louise (Artist), editor.
Title: Art and biotechnology : viral culture from CRISPR to COVID / edited by Claire Nettleton and Louise Mackenzie.
Description: London : Bloomsbury Visual Arts, [2024] | Series: Biotechne : interthinking art, science and design | Includes bibliographical references and index.
Identifiers: LCCN 2023050989 (print) | LCCN 2023050990 (ebook) | ISBN 9781350376021 (paperback) | ISBN 9781350376038 (hardback) | ISBN 9781350376045 (ebook) | ISBN 9781350376052 (pdf)
Subjects: LCSH: Science and the arts. | Art and biology. | Creation (Literary, artistic, etc.)
Classification: LCC NX180.S3 A725 2024 (print) | LCC NX180.S3 (ebook) | DDC 700.1/05—dc23/eng/20240109
LC record available at https://lccn.loc.gov/2023050989
LC ebook record available at https://lccn.loc.gov/2023050990

ISBN: HB: 978-1-3503-7603-8
PB: 978-1-3503-7602-1
ePDF: 978-1-3503-7605-2
eBook: 978-1-3503-7604-5

Series: Biotechne: Interthinking Art, Science and Design

Typeset by RefineCatch Limited, Bungay, Suffolk
Printed and bound in India

To find out more about our authors and books visit www.bloomsbury.com and sign up for our newsletters.

To our families

CONTENTS

PLATES

FIGURES

NOTES ON CONTRIBUTORS

Pablo Baler is a novelist and professor of Latin American Literature at California State University, Los Angeles. His research interest focuses on the intersection of Visual Art, Literature, and Philosophy. Among other books, he is the author of the novels *Circa* (1999), *Chabrancán* (2020), and *El lejano desoriente* (2022), as well as the essay *Latin-American Neo-Baroque: Senses of Distortion* (2016). Baler is also the editor of *The Next Thing: Art in the Twenty-First Century* (2013).

Roberta Buiani is program coordinator of the NewONE, an interdisciplinary experiential learning program at New College, University of Toronto, and co-founder and director of ArtSci Salon, a program connecting artistic and scientific practices through research-creation, based at the Fields Institute for Research in Mathematical Sciences.

Paul Cahill is an associate professor of Spanish at Pomona College. His research interests focus on twentieth- and twenty-first-century Spanish poetry and he has published articles and book chapters about the work of Jenaro Talens, Ana Rossetti, Concha García, and Antonio Méndez Rubio. In 2013 he prepared a critical edition of two books of poetry by Jenaro Talens, *Tabula rasa* and *El sueño del origen y la muerte*.

Oron Catts is the director of SymbioticA, The University of Western Australia. Oron's interest is *Life*, more specifically the shifting relations and perceptions of life in the light of new knowledge and its applications. Often working in collaboration with other artists and scientists, he has developed a body of work that speaks volumes about the need for new cultural articulations of evolving concepts of life.

Joe Davis is a research affiliate at MIT's Department of Biology and Church Laboratory at Harvard Medical School. His pioneering art-science work includes creating "DNA programming languages" for inserting texts and images into DNA. Notable works include *Earth Sphere*, *RuBis Stars*, and *Microvenus*. His *Bacterial Radio* received the Golden Nica prize (2012) and his collaborative project *Baitul Ma'mur* with Pakistani artist Sarah Khan received an Honorary Mention (2021) at Prix Ars Electronica.

Anna Dumitriu is an award-winning internationally renowned British artist who works with BioArt, sculpture, installation, and digital media to explore our relationship to infectious diseases, synthetic biology, and robotics. Past exhibitions include ZKM, Ars Electronica, BOZAR, The Picasso Museum, HeK Basel, Science Gallery Detroit, MOCA Taipei, LABoral, Art Laboratory Berlin, and Eden Project. She holds visiting research fellowships at the University of Hertfordshire, and Brighton and Sussex Medical School, as well as artist-in-residence roles with the Modernising Medical Microbiology Project at the University of Oxford, and the National Collection of Type Cultures at The UK Health Security Agency. https://annadumitriu.co.uk/

Ira Fleming is an MD/PhD student at the University of Colorado in the Department of Molecular Biology. His primary research interests include the immunology of infection, bioinformatics, and RNA biology.

Kathy High is professor in the Arts Department at Rensselaer Polytechnic Institute, New York. She is an artist, filmmaker, and educator who collaborates with scientists and activists, and considers living systems, and animal sentience, and the ethical dilemmas of biotechnology and medical industries. She is the coordinator of the environmental justice community science project NATURE Lab with The Sanctuary for Independent Media. She is committed to queer and feminist approaches to community DIY science, and learning-by-doing. https://www.mediasanctuary.org/initiatives/nature-lab/

Dalila Honorato is a tenured associate professor in Aesthetics and Visual Semiotics, Ionian University, Greece, and collaborator at the Center of Philosophy of Sciences, University of Lisbon, Portugal. Co-founder of the Interactive Arts Lab, she is the head of the interdisciplinary conference "Taboo-Transgression-Transcendence in Art & Science," and, together with Marta de Menezes, the conceptualizer and developer of the project "FEMeeting: Women in Art, Science and Technology." https://inarts.eu/en/lab/staff/honorato/

WhiteFeather Hunter is an internationally recognized Canadian artist and scholar, currently completing a PhD in Biological Art at The University of Western Australia (UWA). Her PhD research, entitled *The Witch in the Lab Coat*, is generously supported by a Social Sciences and Humanities Research Council of Canada (SSHRC) Doctoral Fellowship, Australian Government International RTP Scholarship and UWA International Postgraduate Award. Her biotechnological art practice intersects technofeminism, witchcraft, micro- and cellular biology with performance, new media, and craft.

Pei-Ying Lin is a Taiwanese artist and a PhD researcher at Eindhoven University Technology. Her research interests lie within manipulating the boundary of invisible/visible, living/non-living, and finding ways to build tools and methods that facilitate such explorations. She has won awards including Ars Electronica, STARTS Prize, Core77 Awards, BADAward, Tung Chung Prize, and Mingler Scholarship. Her project PSX Consultancy is a permanent collection of the Museum of Architecture and Design, Slovenia.

Christian de Lutz, originally from New York, is a co-founder and co-director of Art Laboratory Berlin. His curatorial work includes numerous exhibitions, workshops, seminars, and publications, focusing on the interface of art, science, and technology in the twenty-first century. His interest is in building multidisciplinary networks and unleashing their creative potential, with special attention given to BioArt and DIY Science initiatives. One of many art-science collaborations was *Mind the Fungi*, collaboration with biotechnologists of TU Berlin.

Louise Mackenzie is an interdisciplinary artist, curator, and writer. Her collaborative practice is concerned with articulating human/material relationships through process, chance, appropriation, and translation. She is a director of ASCUS Art and Science, Edinburgh, lecturer at Duncan of Jordanstone College of Art and Design, Dundee, and artist researcher at both Newcastle and Northumbria Universities in the UK. Her artworks have been exhibited nationally and internationally, including ZKM, Germany and BALTIC, UK, and her most recent publications include *Bioprotopia* (2023) and *Instructions for Non-Human Listening* (2023).

Molly McKinney is an independent researcher and registered nurse. They hold an MSc in public health and are a member of the American Nurses Association, the American Holistic Nurses Association, and the Emergency Nurses Association. Their work draws from multiple academic theories as well as personal experience and is grounded in the material experience of the body. Previous publications include articles in the *Journal of Educational Policies and Current Practices*, *Transgressing Feminist Theory and Discourse* (2018).

Marta de Menezes is a Portuguese artist, with a degree in Fine Arts from the University of Lisbon and an MSt from the University of Oxford. She is director of Cultivamos Cultura, the leading institution devoted to experimental art in Portugal and Ectopia, dedicated to facilitating collaborative work between artists and scientists. Marta de Menezes has worked in the intersection of art and biology since the late 1990s, in the UK, Australia, the Netherlands, and Portugal, exploring the conceptual and aesthetic opportunities offered by biological sciences for visual representation in the arts. http://martademenezes.com

Claire Correo Nettleton is the academic curator at the Benton Museum of Art at Pomona College and adjunct professor of Art History at Harvey Mudd College. Specializing in collaborations between science, art, and literature from the nineteenth century to today, she has co-organized colloquia and exhibitions including *Viral Culture* (2018), the *Benton Museum Art-Science Colloquium Series* (2021–), and *Parisian Ecologies* (2022). She is author of numerous publications including *The Artist as Animal in Nineteenth-Century French Literature* (2019).

Regine Rapp is an art historian, curator, and co-director of Art Laboratory Berlin. She researches, curates, teaches, and publishes on twenty-first-century art at the interface of science and technology. Next to numerous exhibition series and publications she conceived many international conferences, such as "Synaesthesia. Discussing a Phenomenon in the Arts, Humanities and (Neuro)Science" (2013), "Nonhuman Agents" (2017), "Under the Viral Shadow. Networks in the Age of Technoscience and Infection" (2021), and "Matter of Flux" (2023).

Ken Rinaldo is professor emeritus in the Department of Art, Ohio State University and internationally recognized for bio-robotic art installations. He invents hybrid ecologies with animals, plants, machines, and microbes constructing biological and electromechanical symbionts. Featured in hundreds of books and reviews, his installations have traveled to thirty-five countries. Rinaldo received an Award of Distinction at Ars Electronica, first prize at Vida 3.0, and awards from the UnitedNations, Nuit Blanche, Vancouver Olympics, Kiasma, and Te Papa Museums.

Stephanie Rothenberg is professor in the Department of Art at the University at Buffalo. Rothenberg's interdisciplinary art draws from digital culture, science, and economics. Her exhibitions include ISEA, Eyebeam Art and Technology Center, Sundance Film Festival, LABoral, Transmediale and ZKM Center for Art & Media, and the collection of the Whitney Museum of American Art.

Dolores Steinman is a former senior research associate at the University of Toronto, having had to retire early due to long COVID. She studied the rapport and connection between medical imagery and its nonscientific counterparts. Originally trained as a pediatrician and cell biologist, and then a volunteer docent at the Art Gallery of Ontario, she is driven by her keen interest in placing technology-based medical explorations in the larger context of the arts and humanities.

Charissa N. Terranova researches complex biological systems from a cultural purview, focusing on the history of evolutionary theory, biology, and biocentrism in art, architecture, and design. She is professor of Art and Architectural History at the University of Texas at Dallas. Her next book, *Organic Modernism: from the British Bauhaus to Cybernetics*, is a transdisciplinary study of "organicism," the holistic idea of maintaining the whole is greater than the sum of its parts.

Meredith Tromble is visiting artist, Feminist Research Institute, University of California, Davis, and professor emeritus, Interdisciplinary Studies/Art & Technology, San Francisco Art Institute. Her research interests include collaborative creative practice, ways of knowing, and interspecies communication. She edited *The Art and Films of Lynn Hershman: Secret Agents, Private I* (2005) and co-edited *The Routledge Companion to Biology in Art and Architecture* (2017). Her most recent volume is *Public #59: Interspecies Communication* (2019).

Paul Vanouse is a professor of Art at the University at Buffalo and founding director of the Coalesce Center for Biological Art. His cinema, biological experiments, and interactive installations have been exhibited in venues in over twenty-five countries, including the Louvre and the New Museum. His art received funding from the National Endowment for the Arts (2019) and other organizations. He received a Golden Nica and two Awards of Distinction at Prix Ars Electronica (2019, 2017, 2013).

Ionat Zurr is an associate professor at the School of Design, The University of Western Australia. She is considered a pioneer in the field of Biological Arts. Zurr's ideas and projects reach beyond the confines of art; her work is often cited as inspiration to diverse areas such as new materials, textiles, design, architecture, ethics, fiction, and food.

Karolina Żyniewicz is an internationally recognized artist and researcher. She graduated from the Academy of Fine Arts in Łódź, Department of Visual Arts in 2009, and she earned her PhD in Cultural Sciences from the Nature-Culture Transdisciplinary PhD Program at Artes Liberales Faculty, University of Warsaw in 2023. She calls herself a liminal being because her work is "in-between" art, biotechnology, humanities, and anthropology. Żyniewicz sees her liminal activity as situated knowledge production. http://www.karolinazyniewicz.eu/

ACKNOWLEDGMENTS

We wish to thank our families for their support and encouragement throughout this project: Justin, Henry, Alicia, and Genevieve Slosky, Susan Correo, Craig Nettleton, and Damian, Lewis, Oscar, Benedict, and Joseph Murphy, Helen and Jim Macgregor, Kenneth and Rose Mackenzie, and Helen Murphy. We also wish to thank series editors Charissa Terranova and Meredith Tromble and the team at Bloomsbury for their collaboration and inspiration. We would like to thank Pomona College for supporting this book project and the Viral Culture colloquium and Frankenstein Bicentennial through a large research grant. We would like to give special thanks to Janet Benton, Chair of the Board of Trustees, and Gabrielle Starr, President of the College, for their support of interdisciplinary work. We also thank Pierre Englebert for his support of this final publication production with a small faculty research grant. We would further like to thank Carmen McLeod and Martyn Dade-Robertson at the Hub for Biotechnology in the Built Environment, Newcastle University, and Kirsty Macari and Pernille Spence at Duncan of Jordanstone College of Art and Design, Dundee University, for supporting Mackenzie's involvement in this project. In addition, Claire would like to offer thanks to the Benton Museum of Art and its director Victoria Sancho Lobis for her unwavering support, her interest in art-science, and the development of the Benton Art-Science colloquium series. She would like to acknowledge Pomona College students Iren Coskun '21, Ira Fleming '18, Franco Liu '20, Rena Hernandez '20, Scott Pease '19, Lilly Thomey '19, and professors Rachel Mayeri at Harvey Mudd College and Andre Calvacanti at Pomona College for their invaluable insight and assistance with the Viral Culture Colloquium. A special thanks to Ira for his book revisions regarding CRISPR. Claire would also like to thank the Pomona College Department of Romance Languages and Literatures professors Jack Abecassis, Paul Cahill, Virginie Duzer, José Cartagena-Calderón, and Margaret Waller for their help and collaboration. Claire would like to give special thanks to Joe Davis and Seth Shipman for spending time with her at Church Lab and for inspiring this project, and François-Joseph Lapointe and Marion Cloutier at the Université de Montréal for their collaboration with the Viral Culture Colloquium. Claire would also like to thank Jennifer Doudna, Nobel Laureate and Pomona College trustee, for their discussion and her inspirational breakthroughs.

Finally, we wish to give thanks to all of our incredible contributors for being so generous with their time and patience as we crafted this project from our initial conversations together during lockdown into what it has become today: a considered reflection on a defining moment in human history and the scientific and cultural elements that both led toward, and continue to relate to, this moment.

Introduction

CRISPR, COVID, Creativity, and Control

Claire Correo Nettleton and Louise Mackenzie

As society attempts to turn the page on the COVID-19 pandemic, which caused nearly 7 million recorded deaths at the time of writing, can life ever truly return to its previous state? Our concepts of the self, the other, and even the definition of humanity have been altered by our understanding of and relationship to microbes and genetics. Whether these relations are conceived of as wars, tragedies, scientific experiments, or simply a part of ourselves, they have ruptured our political, economic, familial, cultural, and philosophical structures. It is for this reason that we offer an interdisciplinary approach to understanding the pandemic, as well as innovations in biotechnology in the last few decades, including genome editing and the creation of vaccines with messenger RNA (mRNA). We ask: Where do the arts, which have been at the forefront of technological innovation, sit in relation to our ongoing viral and biomedical entanglement?

Furthermore, how have the arts shaped discussion around CRISPR genome editing and COVID-19? How have the pandemic and biotechnological innovation transformed the arts and humanities and what it means to be human? To answer these questions, we gathered interdisciplinary artists and scholars to examine the relationship between biotechnological, artistic, and literary innovation. This anthology is urgent in its topicality, embodying the zeitgeist of contemporary nonhuman-to-human viral transmission and gene editing technologies. The volume focuses in particular on how newfound molecular technologies and knowledge

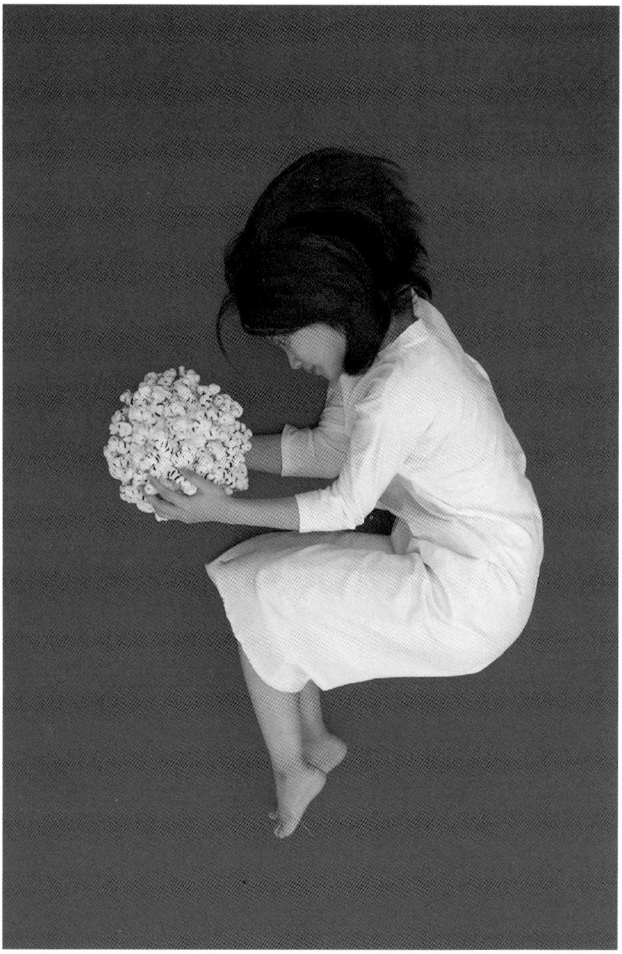

FIGURE 0.1 *Pei-Ying Lin.* Tame is to Tame. *2016.*

regimes, such as CRISPR-Cas9 genome editing and the COVID-19 pandemic, alter understandings of the human experience.

Each of the contributors has an interest or background in biology, blurring the ever-effacing boundary between science and society. Their reflections thus complement research conducted by the medical community—especially those who have documented their personal encounters with COVID-19 (Steinman, Hunter, and McKinney) or even have engaged directly with the SARS-CoV-2 virus in their practice (Dumitriu). Artists are particularly suited to respond to questions of making the seemingly invisible (or microscopic) visible (even sense-able), contributing to public understanding of health. The manipulation of genetic sequences—CRISPR genome editing, contemporary experiments and artworks that embed image, text, and film (Davis, Mackenzie, Nettleton) into DNA—has transformed many narratives, both artistic and scientific, and is thus inextricably tied to the arts and letters. The humanities are a means to think critically and ethically about emerging biotechnologies, especially in the fields of genetics (Mackenzie, Nettleton, Rothenberg) and disease control (Żyniewicz, Catts and Zurr, Vanouse) and their societal impact at a time of rapid acceleration (Terranova). In an era of global heating and mass extinction, biotechnological arts and biodesign can offer both hope for ecological restoration and a warning against the technological dominance of the planet (Nettleton, Rinaldo, Rothenberg, Catts and Zurr). The World Health Organization announced on May 5, 2023, that the COVID-19 pandemic is no longer a global emergency—though people continue to die of the disease around the world. How can art help us process severe racial and economic disparities surrounding the pandemic and the unthinkable loss of millions of people (Buiani, High, Vanouse)?

Viral Cultures: A History

On March 11, 2020, the barriers between laboratory and society dissolved as the World Health Organization declared COVID-19 a pandemic, prompting many countries to initiate social distancing measures and "lockdowns." As COVID-19 closed the old world of carefree physical communication, a new era of virtual interaction opened between networks

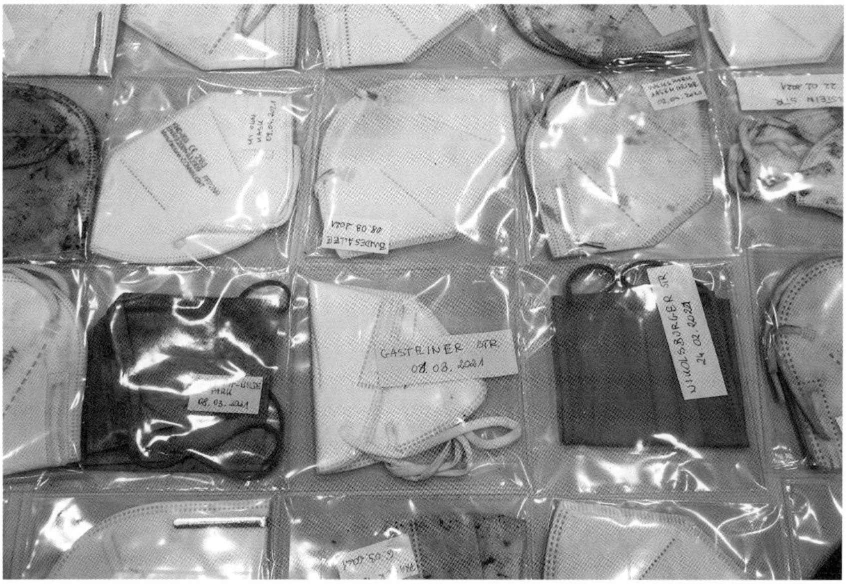

FIGURE 0.2 *Karolina Żyniewicz.* Signs of the Times. *2021.*

and teams. Viral Cultures, a worldwide virtual network of interdisciplinary artists and scholars, amassed over 180 members who met weekly and exchanged views on emerging topics in response to the current world situation, including ways the virus forced creative solutions that blurred the lines between art, humanities, and medicine. Together, we sought a deeper understanding of how viruses operate and their threat to the social order. As founding members of this group, we decided to assemble this collection of essays from a number of these artists and humanists because we believe their perspective on human-viral entanglements has tremendous value today.

Our group Viral Cultures was itself named after a 2018 Pomona College symposium on art, genome editing, and the microbiome. Organized by Claire Nettleton of Pomona College and Rachel Mayeri of Harvey Mudd College, the conference examined biomedical advancement and its relation to artistic and literary innovation. This focus was inspired by the research of Jennifer Doudna, a Pomona College trustee and alumna who received the 2020 Nobel Prize for Chemistry together with Emmanuelle Charpentier for their application of the CRISPR system to healing genetic disease. Specifically, the symposium examined the ways in which CRISPR-Cas9 genome editing—which uses viral DNA present in bacteria to act as scissors to remove target genetic mutations—and understandings of the microbiome, the trillions of microorganisms in the body, alter conceptions of what it means to be human. As conference contributor Lilly Thomey (2018) explained, the microbiome shifts the focus beyond a unique human subject separate from and above all other creatures to humans, bacteria, archaea, fungi, protists and viruses existing as an interdependent colony (Vanouse, Rinaldo, Catts and Zurr).

CRISPR (clustered regularly interspaced short palindromic repeats) is a naturally occurring immune system within bacteria and archaea, which relies upon palindromic sequences of nucleotides in DNA plus snippets of viral DNA, which give the cell the ability to recognize and target foreign viral invaders for destruction by the Cas9 protein. In July 2023, a paper published in *Nature* reported that eukaryotes (including humans) are equipped with an enzyme called Fanzor, which, like Cas9, can be targeted by and toward specific genetic sequences and can be programmed for human genome engineering (Saito, Xu, and Faure et al. 2023). CRISPR-Cas9 allows for the editing of specific sequences of nucleotides—which can modify traits across generations of organisms—a process which shares parallels with writing literary texts. George Church, a geneticist who leads Church Lab at Harvard Medical School and who is responsible for the Personal Genome Project, considers himself a writer and editor of the genome, arguing that genetic code was the very first "text," containing the letters of the four nucleobases A C G T, which transform in RNA (Church and Regis 2014).

In engaging scientists in dialogue with creatives, the symposium warned against potential unregulated and unethical practices, almost foretelling the notorious event of a few months later when Dr. He Jiankui announced to the world in November 2018 that he had created the first genetically modified twin babies, Lulu and Nana, breaching ethical and scientific norms. While CRISPR may potentially treat a myriad of genetic diseases such as cystic fibrosis and Huntington's disease, serious ethical questions also occur as to the long-term consequences of altering the heritable genome (Greely 2019), as well as to what constitutes a disease or genetic error.

The decreasing cost and increasing simplicity of CRISPR now make efficient and precise genome modification technology widely available, opening the door for exploration of this field beyond scientific discovery alone.

As architects Morrow, Bridgens, and artist Mackenzie point out, "biotechnology, biological engineering, synthetic biology, and biofabrication (there are many names emerging for the process of crafting with living material) are no longer the sole preserve of bioscientists with specialist skills" (Morrow et al. 2023). The future of fundamental human needs such as shelter, food, and clothing are drawn in, as the fields of architecture, agriculture, and the food and fashion industries explore the potential of genetic technologies through the increasingly popular catch-all term, "biodesign." Within experimental architecture, genetic approaches are being employed to develop responsive living building materials that can (for example) grow and harden into building blocks (Arnardottir 2023). The genetic modification of crops to develop faster-growing or longer-lasting produce (Zhang et al. 2016), to provide drought-resistance (Khan et al. 2019), or to resist pesticides (Seralini 2020) has swept through the farming and agricultural industries globally. In tandem, the ethical imperative to reduce consumption of meat in the face of

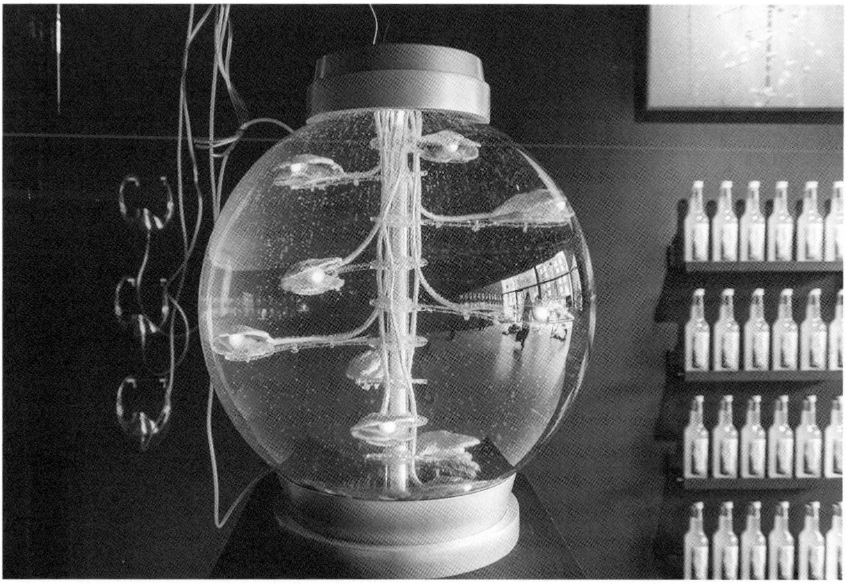

FIGURE 0.3 *Stephanie Rothenberg.* Aquadisia. *2022.*

global heating has seen a surge in the development of both meat-free burgers and vegan leather, using alternative sources that can include the genetic redesign of plant or microbial material.

With CRISPR-mediated germline editing, ethical concerns also arise such as the potential for experimentation in eugenics and the creation of "designer" babies or the possible environmental impact of CRISPR-modified species like Zika virus-resistant mosquitos. In recent years, scientists have created animal hybrids that have rivaled medieval bestiaries, including catfish with alligator DNA (Technology Review 2023), a rat-mouse, and human-pig embryos in view of creating chimeras for organ transplants (Liu et al. 2016). Given the creation of "glowing animals" over the last few decades (Combs 2009), philosopher Vilém Flusser's portent of horses radiating "phosphorescent colours over the nocturnal shadows of the land" (1988) is altogether more plausible. Scholars in the arts, humanities, and social sciences can serve to nourish, support, evaluate, and sometimes warn against technological innovation by offering a historical and cultural framework for understanding its impact on society.

Human knowledge of DNA has been inextricably bound to viruses since the late twentieth century. Some of the earliest research into genetics can be traced to British army surgeon Frederick Griffith's work during the deadly flu epidemic of 1918 (Griffith 1928), which contributed to the understanding that the component responsible for heredity within a cell is DNA, confirmed in further research on pneumococcal infections by Oswald Avery, Colin MacLeod, and Maclyn McCarty (1944). Shortly after, Alfred Hershey and Martha Chase (1952), working on a virus that infects bacteria, conclusively proved DNA's role in heredity. Fast-forward to the present day and, as the result of the COVID-19 global viral pandemic, the relationship between humanity, DNA, and viruses is once again at the forefront of scientific and cultural research.

The invention of CRISPR genome editing and the COVID-19 pandemic are two world-changing biological events that have occurred in the last decade, altering individuals, relationships, societies, and ecologies. As attested to in the essays in this volume, these twin figures of biotechnological advance and microbiological potency can be linked to the practices of artists both before and since their genesis. The two directly intertwine in rapid COVID tests that use CRISPR enzymes to detect the virus (Fisher 2020). Genetic technology was comprehensively deployed in the rapid development of COVID-19 vaccinations across the globe. The Moderna and Pfizer-BioNTech mRNA vaccines demonstrate the tremendous implications of synthetic biology for both future medical research and the functioning of society. Humanity, once skeptical of the term GMO, now comes face to face (or needle to arm) with genetically engineered vaccines that aim to protect us from the threat of viral infection. This volume thus fosters a critical engagement between the arts and humanities and synthetic biology and biomedical advancement.

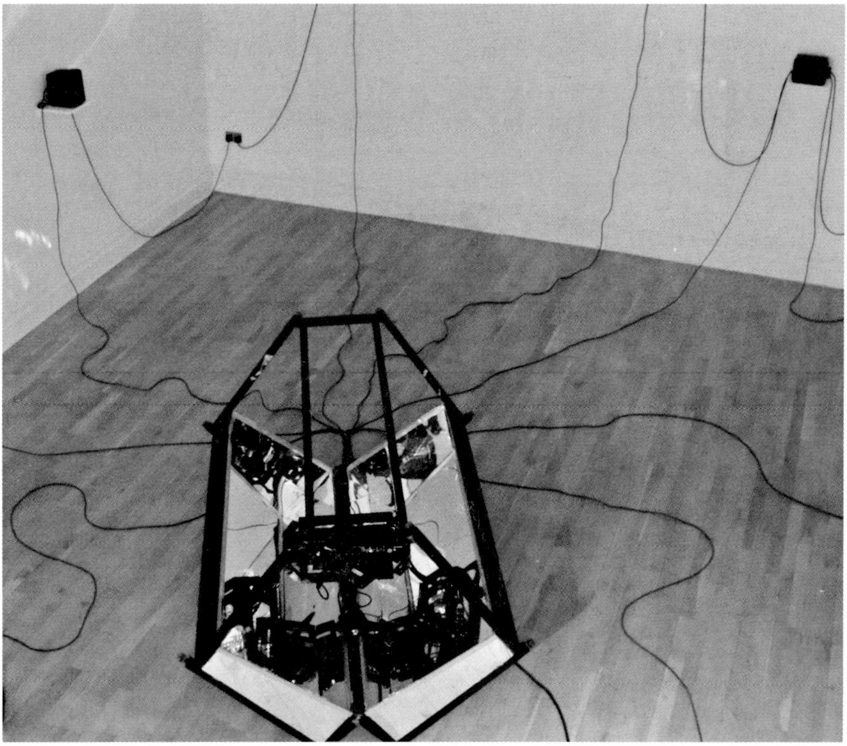

FIGURE 0.4 *Louise Mackenzie*. -Phage. *2017*.

Organization

This book is organized in two parts: 1. Biotechnology and the Arts: Studies and Practice, and 2. COVID-19 and the Arts: Reflections and Practice. Just as innovations in biotechnology and genomics intersect with COVID-19, the subcategories between studies, reflections, and craft are fluid. They draw from each other and the practices of other artists, scientists, and humanists and from the nonhumans that exist inside and around us.

Biotechnology and the Arts

In this section, we discuss transformations in arts, humanities, and humankind in light of CRISPR and related genetic technologies. We engage in a critical reflection on genetic manipulation, especially CRISPR species hybrids and deextinction efforts, which also merit a thorough ethical examination (Nettleton). Claire Correo Nettleton's essay examines a Church Lab experiment which encodes Eadweard Muybridge's *Plate 626* (1887), a film of a galloping horse ridden by an African American jockey, into bacterial DNA using

CRISPR. The author contextualizes these experiments within the arts and humanities, from the invention of cinema to Jordan Peele's film *Nope* (2022). She argues that CRISPR research intersects with cinema and animal studies through their interest in representing, resurrecting, and recreating animals.

Broadening the context to art's relationship with contemporary biotechnology, scholars Charissa Terranova and Pablo Baler introduce us to the live and even self-generating nature of art that engages with biology. Terranova's essay casts autonomy and agency through the lens of autopoiesis in the work of Charles Darwin. Autopoiesis or "self-creation" is a defining quality of the work of bio-artists who recognize the creative agency of nonhuman organisms and their symbiotic relationship with microbes. Bioart can be a means of deacceleration (Terranova), which resists categorization, conceptualization, and traditional structures of power (Baler). Integrating the blurred division between species in the writings of Giorgio Agamben, Pablo Baler explores the meaning and the impact of some current tendencies in bioart. Beginning with Luis Buñuel's 1962 Mexican surrealist film *El ángel exterminador* (The Exterminating Angel), Baler's text reveals the boundary-crossing nature of contemporary art-science practices, which often express the interplay between power and resistance.

Through music, poetry, and art, contributors Paul Cahill, Joe Davis, Ira Fleming, and Louise Mackenzie reflect upon genetics as creative code. Cahill's essay illustrates the parallels between visual poetry and genetic manipulation through the poetry of Spanish writer Juan Eduardo Cirlot (1916–73) whose work, beginning in the 1950s, explored permutation as a poetic technique. DNA can also resemble and be paired with musical notes, which may help us better understand genetic composition (Fleming). Fleming's own method of gene-to-music translation involves the assignment of amino acids to notes and the incorporation of consonance or dissonance based on the degree of gene conservation across species. The resulting curious melody draws to attention the importance of form, function, and meaning in both biology and music.

While Cahill and Fleming present studies on what translation can offer to knowledge-making, artists Joe Davis and Louise Mackenzie practice genetic translation as a poetic and conceptual tool within broader cultural themes. In 1986, the original "DNA poet," Joe Davis, resident artist-scientist in Church Lab, pioneered a method for encoding an image into the genes of live bacteria in the artwork *Microvenus* (1986). Davis's most recent art-science project, undertaken during quarantine in 2020 with collaborators including artist-scientist Sarah Khan in Peshawar, Pakistan, draws upon patterns in Islamic art to present a novel means for encoding information within DNA, while also delivering a cross-cultural message of hope. Invoking the tradition of making angels by uttering the Arabic phrase, "Subhan Allah," Davis and collaborators conceive the presentation of more than 200 quadrillion angels on the head of a pin in poetic counterbalance to the tragic number of deaths due to the COVID-19 pandemic.

The arts can bring to our attention issues in scientific technology and public health through critical and speculative engagement with genetic technologies and nonhuman species (Mackenzie, Rothenberg, Rinaldo). Mackenzie's *Blue Green Planet*, an image that is simultaneously a globe and a living mass of cyanobacteria, highlights the importance of microbial life to the production of oxygen in the earth's atmosphere. Like Davis, artist Louise Mackenzie also translates information into DNA. In her *Pithos* series of works (2016–), her message is a question to the microorganisms in receipt of this DNA, from whom she waits endlessly for an answer, highlighting our inability to fully comprehend other forms of life. Mackenzie's 2017 work *-Phage* (Figure 0.4) alludes to the genetic manipulation that she was practicing in the laboratory: cutting and splicing images and audio in a sculptural simulation of a viral organism. Through public workshops and film, Mackenzie examines the role of performative art in considering human/nonhuman ethical relations regarding genetic modification. Introducing anthropomorphism as a performative technique for genetic modification practices, Mackenzie considers microorganisms as collectively complex forms, possessed of nonhuman sentience, allowing us to experience genetic modification in a relational context both within and beyond the laboratory.

Further speculation on genetic futures can be found in chapters by Ken Rinaldo and Stephanie Rothenberg, who through their artworks lead us into the tantalizing possibilities that CRISPR futures may hold, while also alerting us to the associated ethical complexities. Rinaldo's multidisciplinary project *CRISPR Seed Resurrection* (2021) explores how our use of CRISPR technology in seeds might contribute to the survival of our species and others against climate change, while Rothenberg's *Aphrodisiac in the Machine* imagines a new species of CRISPR genetically modified oysters, which convert polluted waters into an aphrodisia-inducing liquid. Rothenberg's work is environmental science fiction displayed through installations, videos, and performances. Building on the libidinous myth of the oyster, the narrative draws from the writings of black feminist writer Audre Lorde and her notion of the erotic as a redefined power of feeling, which challenges patriarchal notions of the erotic as a commodified good. This fantastical project aims to critique some of the ethical contradictions of gene modification and provide a space for envisioning alternative multispecies future-making.

As Viral Cultures participants Iren Coskun, creator of the interdisciplinary art project *BioMap* (2019), and Thomey stated (2021), "in the times of COVID-19, our viral awareness and cultural connection to our microbial communities has shifted, inciting fear of the little things we cannot see and do not fully understand." Rinaldo's work pushes us to explore our symbiotic relationship to microbial material and changing notions of life in the age of CRISPR and COVID-19. Exhibited in January 2020, as COVID-19 began to spread globally, Rinaldo's *Borderless Bacteria/Colonialist Cash* (2020) features currencies from countries in conflict. The bills, presented in culture-rich Petri dishes, grow bacteria, fungi, and viruses that have traveled from our hands to their surfaces, making manifest biopolitics at the micro scale.

COVID-19 and the Arts

The next section focuses on interdisciplinary engagement with the COVID-19 pandemic and its impact on humanity. Artist Meredith Tromble explores the potential of metaphors developed from research into evolution of the epigenome (the numerous chemical compounds that can inform the genome what to do) to generate new understandings about pandemic-induced change in the art world. Roberta Buiani gives examples in the arts, which help us reflect on the possibility of embracing the viral at a time of complexity, risk, and polarization. In their discussion of Art Laboratory Berlin's exhibitions and programming, Regine Rapp and Christian de Lutz reveal the ways in which the pandemic has disrupted and recreated forms

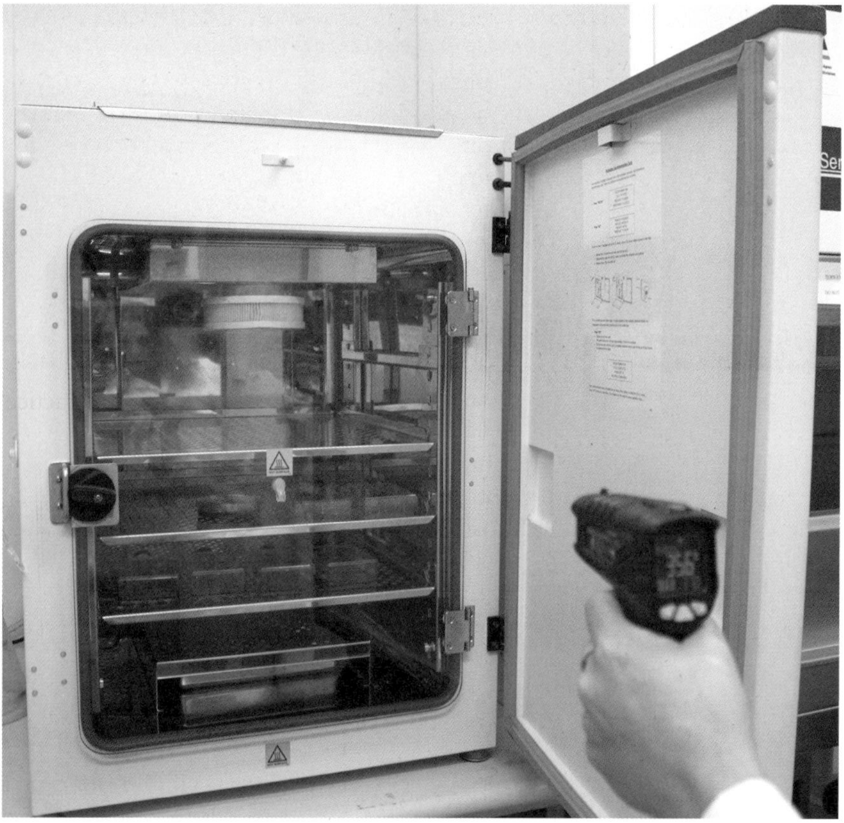

FIGURE 0.5 *Oron Catts and Ionat Zurr.* Thermoception. *2020.*

of exchange, analyzing the tension between the ethical obligation to minimize global travel and the human desire to gather. Similarly, Marta de Menezes and Dalila Honorato illustrate the ways in which COVID-19 altered gatherings of the FEMeeting network, which advances research, collaboration, and community in art and science by women (and those identifying as female). In their chronicles of ruptures and new opportunities created during the pandemic, the authors trace the evolution of this metaphorical "antibodies network."

As personal narratives reveal, traditional scientific and medical research could not keep up with the rapid pace of the early stages of the pandemic, causing long COVID symptoms to be dismissed and bodies to be considered disposable, especially in women, minorities, and frontline workers (Steinman, Hunter, and McKinney). Citizen science, including websites which tracked cases to help map the spread of the virus, and logged patients' symptoms, played a key role in filling information gaps (Ziegler et al. 2022) and legitimizing these experiences. COVID-19 has now ushered in a new age of bioliteracy with increasing public interest and debate in the field of epidemiology.

WhiteFeather Hunter, artist, and Molly McKinney, registered nurse, posit a meaty metaphor of the "Turducken" for capitalist consumption at the expense of the undervalued bodies which support it. They support a material feminist analysis with personal narratives from a long COVID patient and a front-line medical worker whose concerns regarding the virus' impact were dismissed. Similarly, as one of the early COVID cases near the European epicenter of Bergamo, Italy, medical doctor and cell biology researcher Dolores Steinman offers a narrative medicine account with long COVID, emphasizing the importance of interdisciplinary online communities in an era plagued by misinformation and biases.

Creativity can be lifesaving. In the early stages of the pandemic, do-it-yourself innovation offered solutions at a time of medical shortages, from homemade masks to makeshift ventilators (Avril 2020), while data visualization bridged the gap between art and science. Biological art practice can reveal larger societal impacts of the pandemic. Referencing both COVID-19 and the genetic technologies that surround it, Oron Catts and Ionat Zurr offer the concept of thermobiopolitics to navigate systems of control in our present viral culture. They trace the history of the thermostat, which becomes the point of departure for their *Compostcubator* (2019) series of artworks that focus on the regulation of life through temperature and the ethics and politics that lie therein—from planetary heating to cross-border temperature checks for viral infection (Figure 0.5). Several essays express the need for a supportive community for women and suggest alternative scientific and health-related paradigms. *Shielding* (Dumitriu 2020) looks at the impact of lockdown on victims of domestic abuse, expressing ethical, aesthetic, and safety concerns around the virus. Artist Anna Dumitriu, with collaborators Ines Moura and Jane Freeman, created this artwork as part of a series which incorporates actual SARS-CoV-2 RNA (coronavirus) from a plasmid construct. This is a safe, non-infectious reagent for SARS-CoV-2 RNA research. The issue of care as the basis of ethical practice in science, found in the ideas of Maria Puig de la Bellacasa, Henry Greely and Anat Pick, is one of several philosophical currents that run through this volume.

Integrating writings by Bhaktin, Paul Vanouse's essay affirms the physical, nonhuman, and nonverbal aspects of speech, in light of the COVID-19 pandemic. Through *Utter*, which is part essay, part in-progress art installation, Paul Vanouse investigates the material component of utterances, including the release of breath, saliva, and microorganisms. The essay also examines historic racial prejudices that have contributed to disparities in illnesses such as cholera. In the age of COVID-19, the connection between racial biases and disease is all the more relevant, as members of racial and ethnic minority groups in the United States including African Americans and Hispanics/Latinos have experienced higher risks of disease positivity, severity, and inpatient mortality (Vardar et al. 2023). From a culture that does not often discuss death, the astronomical number of deaths due to the pandemic, especially in marginalized communities, has placed death very much in the foreground. This inspired artist Kathy High to assemble a *Death Tool Kit*, with input from five artist colleagues whose practices relate closely to the process of dying. High's radical and extremely moving essay helps us grapple with what might be called a "good death," crossing over, grieving, releasing, and transforming to our next selves.

In contrast, Karolina Żyniewicz discusses how COVID-19 creates the sense of being in a global Panopticon, where most feel observed and controlled. Combining theory by Michel Foucault, Jeremy Bentham, and Bruno Latour, and her artwork *Signs of the Times* (Figure 0.2), the author argues that the COVID-19 pandemic creates new understandings of the concept of the Panopticon within the context of a global viral outbreak. As surveillance alters our behavior, Pei-Ying Lin's essay illustrates the ways in which the virus affects human consciousness. She explores this phenomenon through two participatory artworks *Virophilia* and *Discourse of a Viral Boundary*. By engaging in "pandemic storytelling," participants explore unconventional situations regarding human-virus interactions. These art experiences allow participants to contemplate the pandemic from various perspectives and create new understandings of human-viral entanglements.

As we focus attention on global events such as the COVID-19 pandemic and the climate crisis, interdisciplinary subjects of study such as medical humanities and environmental art are an increasingly common listing in university curricula. The seemingly objective sciences are now under the critical eye of scholars in the humanities and art; the study of biology becomes cultural, allowing for new perspectives to shed light on firmly held beliefs and processes. Through COVID-19 we find ourselves in an evolving picture of genetic creativity, where human knowledge of DNA has moved beyond scientific understanding to a creative tool. Genetics begins to more closely reflect art as DNA becomes the cut-up text of scientists rather than Dada poets and assemblages are constructed from chimpanzee adenovirus and genetically modified human kidney cells (*ARCHIVE: Information for UK Recipients on COVID-19 Vaccine AstraZeneca*) as readily as metal, wood, and glass. With growing warnings of future pandemics, we are urged to respond, but how and to what effect?

This series of essays brings into sharp focus the value of artistic practice surrounding our reactions and responses to the COVID-19 pandemic through their reflections on interaction with the virus as well as through artworks that shape our understanding of molecular material and genetic technologies. Although now over three years on from the first declaration of a global pandemic, the impact of coronavirus is still keenly felt around the globe and these first (cultural) responses are no doubt only a fraction of the story that is emerging.

References

Avery, O. T., MacLeod, C. M., and McCarty, M. (1944). "Studies on the chemical nature of the substance inducing transformation of pneumococcal types : induction of transformation by a desoxyribonucleic acid fraction isolated from pneumococcus type III." *The Journal of Experimental Medicine* 79(2): 137. https://doi.org/10.1084/JEM.79.2.13

Avril, T. (2020). "MacGyvering the Coronavirus: Designers Rig Makeshift Ventilators, Valves and More." *The Philadelphia Inquirer*, April 30. https://www.inquirer.com/health/coronavirus/coronavirus-ventilators-shortage-innovation-engineering-design-diy-20200330.html

Arnardottir, T. (2023). "Bacterial Sculpting—Customising Biofabrication Techniques for Biomineralisation." In Morrow, R., Bridgens, B., and Mackenzie, L. (eds.) 2023, *Bioprotopia: Designing the Built Environment with Living Organisms*. Birkhauser.

Church, G., and Regis, E. (2014). *Regenesis: How Synthetic Biology Will Reinvent Nature and Ourselves*. Hachette.

Combs, C. (2009). "Animals Shining for Science." *National Geographic* online. https://www.nationalgeographic.com/animals/article/shining-for-science-pictures-of-glowing-animals

Coskun, I., and Thomey, L. (2021). Correspondence with Claire Nettleton on April 5.

Fisher, C. (2020). "The FDA Approves a Rapid COVID-19 Test that Uses CRISPR." https://www.engadget.com/fda-sherlock-biosciences-covid-19-crispr-test-162943598.html

Flusser, V. (1988). "Curie's Children." *Art Forum* 27(2): 9.

Greely, H (2019). "CRISPR'd Babies: Human Germline Genome Editing in The 'He Jiankui affair'," *J Law Biosci* 13.6(1): 111–83. doi: 10.1093/jlb/lsz010

Griffith, F. (1928). "The Significance of Pneumococcal Types." *The Journal of Hygiene* 27(2): 113. https://doi.org/10.1017/S0022172400031879

Hershey, A. D., and Chase, M. (1952). "Independent functions of viral protein and nucleic acid in growth of bacteriophage." *The Journal of General Physiology* 36(1): 39–56. https://doi.org/10.1085/jgp.36.1.39

Khan, S., Anwar, S., Yu, S., Sun, M., Yang, Z., and Gao, Z. Q. (2019). "Development of Drought-Tolerant Transgenic Wheat: Achievements and Limitations." *International Journal of Molecular Sciences* 20(13). https://doi.org/10.3390/IJMS20133350

Liu, Y., et al. (2016). "Chromatin Features and the Epigenetic Regulation of Pluripotency States in ESCs," *Cell* 164(3): 532–43. doi: 10.1016/j.cell.2015.11.052

Morrow, R., Bridgens, B., and Mackenzie, L. (eds.) 2023. *Bioprotopia: Designing the Built Environment with Living Organisms*. Birkhauser.

Saito, M., Xu, P., Faure, G., et al. (2023). "Fanzor is a eukaryotic programmable RNA-guided endonuclease." *Nature*. doi: 10.1038/s41586-023-06356-2

Seralini, G. E. (2020). "Update on long-term toxicity of agricultural GMOs tolerant to roundup." *Environmental Sciences Europe* 32(1): 1–7. https://doi.org/10.1186/S12302-020-0296-8/METRICS

Technology Review (2023). "CRISPR Used to Insert Alligator Genes into Catfish." *MIT Technology Review*. https://www.technologyreview.com/2023/01/19/1067092/crispr-alligator-gene-catfish/

Thomey, L. (2018). "Introduction to the Microbiome." Viral Culture Conference. April 26–7. Pomona College.

UK Department of Health (2022). "Information for UK Recipients on COVID-19 Vaccine AstraZeneca." *GOV.UK*. https://www.gov.uk/government/publications/regulatory-approval-of-covid-19-vaccine-astrazeneca/information-for-uk-recipients-on-covid-19-vaccine-astrazeneca

Vardar U., et al. (2023). "Racial Disparities in Patients With COVID-19 Infection: A National Inpatient Sample Analysis." *Cureus* 15(2). https://pubmed.ncbi.nlm.nih.gov/36942174/

Zhang, C., Wohlhueter, R., and Zhang, H. (2016). "Genetically modified foods: A critical review of their promise and problems." *Food Science and Human Wellness* 5(3): 116–23.

Ziegler, S., et al. (2022). "Long COVID Citizen Scientists: Developing a Needs-Based Research Agenda by Persons Affected by Long COVID." *Patient* 15(5): 565–76. doi: 10.1007/s40271-022-00579-7

PART ONE

Biotechnology and the Arts: Studies

1

Resurrecting the Woolly Mammoth and Muybridge's Horse

CRISPR, Cinema, and Species Revival

Claire Correo Nettleton

During a lab tour at Harvard Medical School, I followed my guide Joe Davis, the resident artist-scientist at the George Church Laboratory, into a guarded elevator. Before stepping out, Davis paused dramatically and stated with an enormous grin above his white beard, "Welcome to Church Lab." The homage was apt—for a moment, it was as if we had entered *Jurassic Park* (Spielberg 1993). I was offered a peek into a microscope to view Asian elephant DNA, which George Church, a pioneer in the field of genomics, was attempting to edit to resurrect the woolly mammoth (Zimmer 2021). However, my primary focus, as an animal studies scholar now turned museum curator, was not this mammoth project or its *Jurassic Park*-like implications, but rather the resurrection of another creature that also made cinematic history.

In 2016, Harvard University screened a film of a galloping horse. Rather than a theater, the five-frame film aired in Church Lab, which projected the movie after recovering it from encoded bacterial DNA. Featuring an anonymous African American jockey riding a racehorse, the film was derived from Eadweard Muybridge's *Plate 626: Gallop, thoroughbred bay mare Annie G* (Figure 1.2), published in his *Animal Locomotion* series in 1887. In the experiment by scientists Seth Shipman, Jeff Nivala, Jeffrey Macklis,

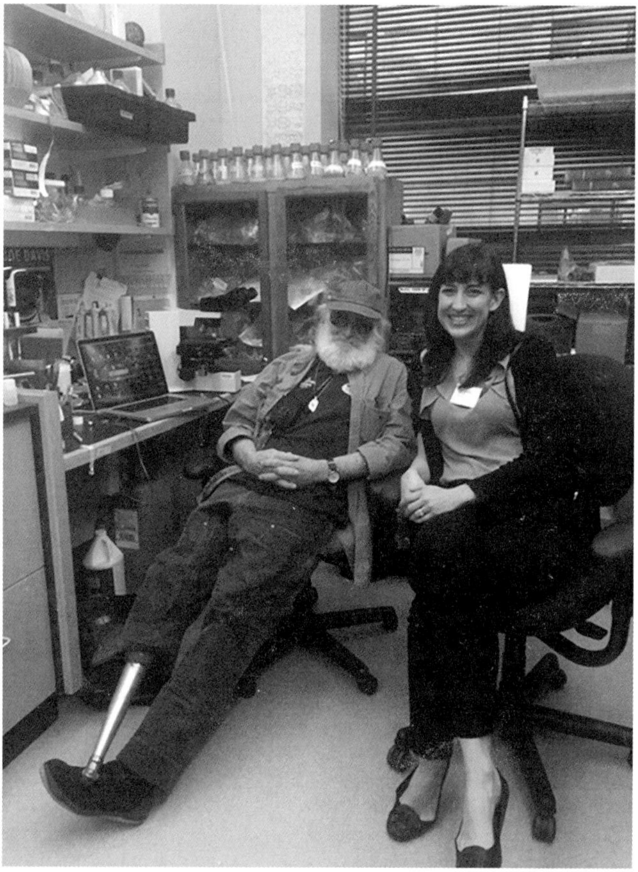

FIGURE 1.1 *Claire Correo Nettleton and Joe Davis at Church Lab, 2018, Claire Correo Nettleton.*

and George Church used CRISPR-Cas, a revolutionary gene editing device adapted from microbial immune systems, to modify the bacteria's genetic structure to encode the film. The technique was inspired by Davis's pioneering art-science work *Microvenus* (1986), the first project to insert images into DNA. Shipman's team reproduced and played the film back from the bacterial offspring with 90-percent accuracy, which reinforced their hopes of using DNA as a durable data storage device and a molecular recorder.

This essay integrates this experiment with discussions with Shipman and gene editing trailblazer Jennifer Doudna, animal studies scholarship, archival research, and cinematic references including Jordan Peele's *Nope* (2022). I argue that three fields—CRISPR research, animal studies, and the history of cinema—intersect due to their interest in representing, resurrecting, and recreating animals. The entanglement of these disciplines raises ethical issues regarding species boundaries and the necessity of human intervention. I ask:

1 What is the relationship between genome editing and cinematic editing, especially concerning the potential to reanimate disappearing animals?
2 Do de-extinction efforts, within the context of this essay, impose human dominance or restore nonhuman life?
3 Could biotechnology revitalize the humanities by breathing new life into works of art, literature, and film? In what ways might a humanistic approach to this experiment and CRISPR allow us to think about ethics?

The fluid relationship between animals, microbes, and humans—as well as science and society—has become all the more relevant since the COVID-19 pandemic with the possibility of viral DNA jumping between species. Vaccines from Moderna and Pfizer-BioNTech demonstrate synthetic biology's implications for medical research and life on this planet.

The 2012 discovery that the CRISPR-Cas immune system in archaea and bacteria was programmable has had revolutionary implications for the treatment of genetic disease and biological research (Mukherjee 2017) but had not received much consideration for nonbiological data storage prior to this experiment (Shipman et al. 2017). While CRISPR-Cas9 is used to make targeted cuts into DNA, Shipman's team utilized the Cas1-Cas2 complex to integrate a moving image into bacterial DNA. In the native bacterial CRISPR-Cas system, DNA from invading viruses is stored within the bacterial genome in sequences referred to as spacers, derived from invading organisms' genetic sequences called protospacers (Shipman et al. 2017).

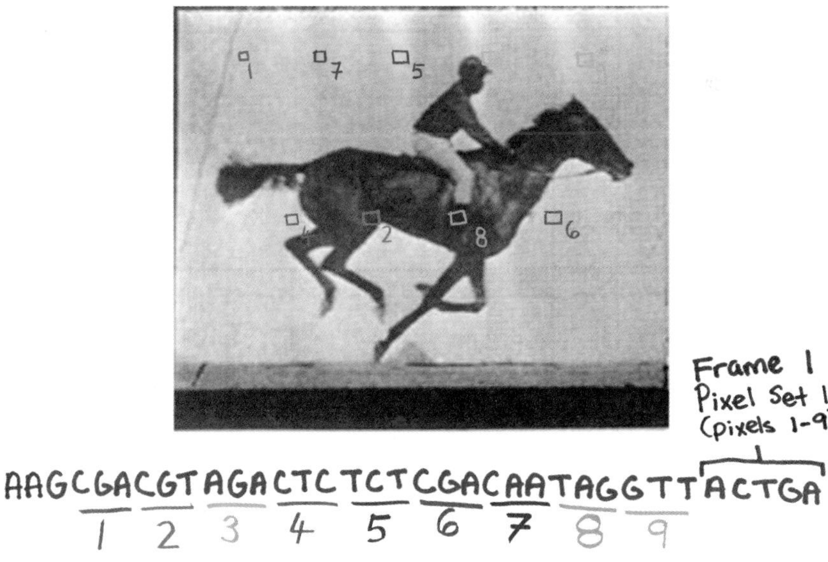

FIGURE 1.2 Encoding memories in Living Cells with CRISPR, *2017*, *The Wyss Institute*.

Shipman's team encoded Muybridge's film using a code (Figure 1.1b), where the shade of each pixel was specified by a nucleotide triplet (Shipman et al. 2017). Shipman's team distributed pixel values in synthetic protospacers, which were transported into *E. coli* that overexpress Cas1 and Cas2, and found that the data was successfully inserted and stored in the genome (Shipman, Nivala, Macklis, and Church 2017). In doing so, the researchers proved that this bacterial immune system can be used to store kilobytes of nonbiological data in the genome of living organisms which can propagate and eventually be sequenced to recall this information. This act pushes the limit of how films and art can be viewed, stored, and reproduced and raises questions about the ethics of inserting data into living beings (see also Chapter 7 in this volume).

In some ways, this parallels Muybridge's breakthroughs on photographing and projecting living creatures. As this anthology argues, CRISPR has implications for not only the sciences but the arts and letters. Shipman's experiment helps redefine humanity and the humanities in the age of CRISPR. Siddhartha Mukherjee writes that our ability to manipulate human genomes alters our conception of being human (Mukherjee 2017: 12). CRISPR establishes new paradigms for viewing and "editing" life, which blur divisions between technological and organic processes and eukaryotic and prokaryotic forms, but with human intervention at its core.

In our interview, Shipman agreed that his team's use of a film of a moving animal points to the potential role of synthetic biology in mitigating mass extinctions. The most famous de-extinction effort is Church Lab's attempted resurrection of the mammoth in collaboration with the company Colossal, the supposed aim of which is restoring ecosystems to combat climate change (Zimmer 2021). While the primary purpose of the racehorse film experiment was to make a prototype molecular recorder in living cells which tracks cellular events (Sample 2017), Shipman informed me that the choice of Muybridge's film also represented historical debates around emergent technologies. He said, "Although the technology is new, the issues rarely are. We scientists are often ill-equipped to explore the historical context." (Shipman, S. (2018). Interview by Claire Nettleton. 12 June). Fittingly, as I will discuss later, the film Shipman used has recently gained attention through its representation in Jordan Peele's movie *Nope* (2022), which exposes the erasure of the jockey's role in Muybridge's film as the first movie star due to the jockey's African American ethnicity. This sci-fi western imagines the descendents of this forgotten jockey attempting to make history by capturing aliens on film, illustrating the danger and the power of new technologies to represent and control life.

In their TEDx talk, Davis and Shipman express the belief that humans will become extinct and the need for data storage beyond human existence (2016). They argue that DNA is one of the most efficient data storage devices. The nucleotide bases A, C, T, and G can be converted to digital information that can last thousands of years (Shipman, Nivala, Macklis, and Church 2017). One way genomics transforms the humanities is through the practices of "reading" human genomes and "writing" living cells through the manipulation and addition of nucleotides (Shipman, Nivala, Macklis, and Church 2017). Church—who in 2021 auctioned off his own genome as an NFT—initiated The Human Genome Project to map the entire genetic sequence of human DNA and was the first to encode a book into DNA (Mukherjee 2017).

1. What is the Relationship between Genome Editing and Cinematic Editing, Especially Regarding the Potential to Reanimate Animals During a Time of their Disappearance?

Contemporary animal studies reevaluates the ways in which we understand animals in art and fiction, arguing that animals are not mere metaphors for humans but are creative entities in their own right. John Berger's "Why Look at Animals?" (1980) argues that art became increasingly focused on animal imagery during a time of unprecedented species extinction due to industrial capitalism (Kolbert 2015), a phenomenon which cinema scholar Akita Lippit argues characterizes modernity itself. Lippit writes that Muybridge's film of a galloping horse coincides with the vanishing of animals in our everyday environment, only to be reborn as specters or the "*undead*" in cinema (2001: 184–5). As animals disappeared, particularly when cars replaced horses, our encounters with fauna often came through stories, images, and moving pictures. The role of art thus transformed into a means of preserving the memory of species. In the era of CRISPR and de-extinction, we must now evaluate this phenomenon within the context of contemporary biotechnology, which not only has the capacity to alter and reconstruct animals and microbes, but also to embed works of art within the DNA of organisms. Lippit writes: "As animals began to disappear from the phenomenal world they became increasingly the subjects of the nineteenth- and twentieth-century reproductive media" (2001: 184–5). The choice of the word "reproductive" (containing the *double entendre* of replicating biologically and technologically) is applicable to Church Lab, which is creating an artificial womb for the embryo of the woolly mammoth. Furthermore, indicated by Berger, especially before the nineteenth century, creatures offered people nonhuman companionship (Nettleton 2019). As Susan McHugh writes, narratives on human-animal companionship are relevant for analyzing biopolitics during an age of mass extinctions (2011: 15). The horse film experiment illustrates the value of interspecies connections while exercising human dominance.

Eadweard Muybridge: From the Zoopraxiscope to *Nope* (2022)

The fact that Shipman's team selected Muybridge's work for this project is especially fitting given that Muybridge himself was an eccentric polymath. A photographer and technological innovator, Muybridge is considered the inventor of both the first film and film projector. The photographer was born in 1830 in what is now Greater London and emigrated to the United States in 1850, living first in New York and then San Francisco (Shimamura 2002: 341). Leland Stanford, Governor of California and racehorse breeder whose breeding farm became the campus of his namesake university, asked Muybridge in 1872 to help him settle an argument on whether any horse had all four legs off the ground while galloping (Brookman, Muybridge, and Braun 2010). While Muybridge's first results were unpublishable due to the lack of sophistication of cameras at the time, he returned to the task in 1877. To capture stills of a racehorse in its full range of motion, Muybridge invented a camera with shutter speeds nearly a hundred times faster than previous cameras, and attached the new cameras to wires that would activate the shutters when the horse tripped them (Brookman et al. 2010). In 1878, Muybridge produced a photographic study, *The Horse in Motion (Sallie Gardner at a Gallop)*, whose contents are considered to constitute the first film when projected as a sequence (Mann 2018). This set of stills could capture what the naked eye or ordinary cameras could not—all four legs of a horse being elevated at once from the ground.

The invention of the motion picture necessitated the creation of the first projector, which Muybridge called the zoopraxiscope (1879), which enabled multiple people to view a moving image at once for the first time (Herbert n.d.). The images were on a rotating glass disc, with lantern light, allowing the pictures to be projected up to ten feet wide. The public could see full-size horses, giving the illusion that animals were stampeding in the room (Herbert n.d.). Muybridge's studies of consecutive motion culminated with his book *Animal Locomotion: An Electro-Photographic Investigation of Consecutive Phases of Animal Movements* (1887), with animals and people in scenes ranging from everyday life to nudes jumping. While the series was criticized at the time for its so-called indecency, current scholarship has analyzed the images in terms of subverting or reinforcing class, gender, or racial norms.

In *Plate 626* (Figure 1.3) (23.5 cm × 30.5 cm), an iteration from Muybridge's horse galloping series which featured sixteen photographs, arranged in four strips, an unnamed African American jockey rides a mare identified as Annie G. Black jockeys were well-represented and respected in the sport at this time, which would prove to be a rarity in repressive and violent Jim Crow America (Hensley 2017). As illustrated in Jordan Peele's film *Nope* (2022), the legacy of these Black jockeys has been all but erased until recent years. Sociologist Catherine Bliss (2020) emphasizes the "need to reconceptualize race in terms of legacies of discrimination, with a deep understanding of discrimination in its many forms." In the film, *Nope*

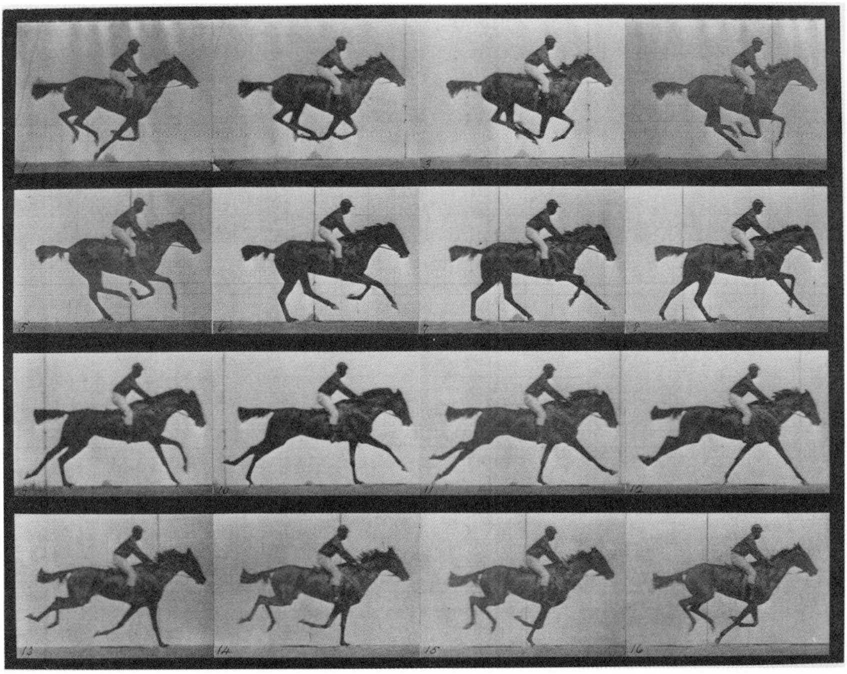

FIGURE 1.3 *Edweard Muybridge,* Animal Locomotion, Plate 626, *Photomechanical Print: Collotype, 1887, Boston Public Library.*

(2022)'s protagonists Emerald Haywood and Otis "O.J." Haywood Jr. are presented as being horse wrangler descendants of the uncredited Black jockey from Muybridge's pioneering series. Early in the film, the two are seen in their house on a horse ranch watching *Plate 626* played as a motion picture. O.J. elucidates that the origin of cinema was an African American on a horse. As Peele (Peele and Palmer 2022) himself states, "We've got the first movie star of all time. And it's a Black man we don't know. We haven't looked. In a lot of ways, the movie became a response to that first film." *Nope* (2022) portrays *Plate 626* and the birth of cinema as a collaborative effort between jockey, horse, and camera operator.

Characters O.J. and Emerald emulate the process of capturing the horse on film as they attempt to document the alien that is hovering above their ranch, using multiple cameras from different angles. They enlist the help of Muybridge-like cinematographer Antlers Holst, who is seen splicing together reels of animal footage with an old-fashioned flatbed analog film editing machine; later in the film, Muybridge proxy Holst is shown using a homemade film camera that does not need electricity. Treating the alien like an animal, O.J.'s skills in communicating with horses, such as not looking them directly in the eye, enable him to break and wrangle the alien like a horse. Holst, whose hubris with his quest to record the alien leads to his own demise, directs O.J., "You break it. I'll get the shot." *Nope* (2022) thus suggests the importance of collaboration in technological and scientific processes while warning against human domination of all life.

Cultural critic Soraya Nadia McDonald voices that Peele's film also recognizes the importance of capturing images in order to be believed (whether by Leland Stanford or by a public that has forgotten the existence of an African American as the very first performer in a motion picture). At the film's climax, a fatigued Emerald spends her last measure of strength to reach an analog camera with a crank, employing a similar development process to Muybridge's, to succeed in capturing photographs of the extraterrestrial (2022). Influenced by *Jurassic Park* (1993), the film expresses mistrust in technology—especially digital technology—as a site of violence (Chow and Zornosa 2022). In the arena of genome engineering, intersectional considerations of CRISPR will hopefully open future avenues of research on the connection between race and contemporary genomics—from the threat of repeating the tragic history of eugenics and cases like the unauthorized use of cells from Henrietta Lacks's body for scientific study, to the politics of ancestry tests (Nelson 2016), and to recent progress in CRISPR correcting sickle cell anemia, a health condition that primarily affects African Americans.

Muybridge and CRISPR

In this section, three relationships are drawn between Muybridge's work and the use of CRISPR technology. The first parallel is the editing of sequences. In the second frame of *Plate 626*, the horse and rider appear to be flying in the air, with all four legs off the ground, poised to return to the ground in the third and fourth frame. Writes biographer Hans Christian Adam, "Muybridge was the first to convert continuous linear movement into a sequence of pictures pieced together from single images" (2010:17). Muybridge also often omitted an image in a sequence, exchanging it for another, which parallels gene editing. In the writings of influential early Soviet filmmaker Sergei Eisenstein, film editing "is frequently likened to a genetic code in which dominant as well as recessive links are made between convergent strings of information" (Lippit 2001: 23–4). Eisenstein's employment of the Kuleshov effect (named for filmmaker Lev Kuleshov, who first demonstrated this phenomenon), in which the perceived meaning of individual images could be manipulated through their juxtaposition with other images, could be compared to the sequencing of DNA nucleotides, where the meaning of DNA (the way DNA is translated into functional proteins by the body) changes depending on the next letter in the sequence.

Fittingly, Muybridge's photographic techniques were used for purposes of selective breeding in animals, enabling owners to select parents for offspring and adjust training regimens to improve critical motion "to correct perceived weaknesses" (Gunning et al. 2003). Doudna argues that CRISPR is a more evolved form of selective breeding, which she claims edits out perceived flaws in plants and animals to create produce such as seedless watermelons, hornless dairy cows, or micropigs (Doudna and Sternberg 2016: 118). The image of a thoroughbred racehorse in Shipman's team's experiment reinforces CRISPR's potential to manipulate livestock.

The third similarity between CRISPR and Muybridge is the use of sequencing to animate animals during the time of their disappearance. Muybridge captures images of American bison, which became nearly extinct in the nineteenth century but have experienced a recent resurgence. When viewing Muybridge's 1887 *Plate 700: Buffalo Galloping* (22.5 cm × 34.8 cm) (Figure 1.4A) animated as a film, with a buffalo charging through the American plains at full force, one notices small inconsistencies between the frames. The buffalo appears larger or smaller or in another location from one second to the next—which demonstrates anomalies in the editing process, revealing that the buffalo is not truly "alive" in front of us. One could make comparisons to the de-extinction process, which does not truly revive lost species, but merely an approximation, where the DNA of extinct species is inserted into that of a close relative, and anomalies can indeed occur.

Muybridge's attempt to reanimate the buffalo is in contrast to photographs of his contemporary William Henry Jackson, whose animal portrait of a buffalo titled *Bull* (Figure 1.3B) (1875), an albumen print on paper (16.51 cm × 10.16 cm), is part of the Benton Museum of Art at Pomona College's collection. Jackson crops the photo in a darker oval shape around a buffalo's head on top of a lighter square background, giving the impression that the buffalo's head is severed and mounted on the left side of the frame as if on a mantle. The animal's still, glassy eyes also seem to be works of taxidermy. Jackson, famous for his explorations of the United States, as well as renegade

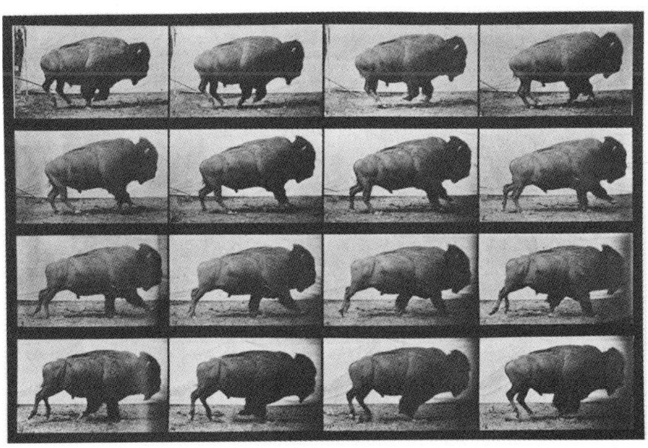

FIGURE 1.4A *Edweard Muybridge,* Animal Locomotion, Plate 700, *Photomechanical Print: Collotype, 1887, Boston Public Library.*

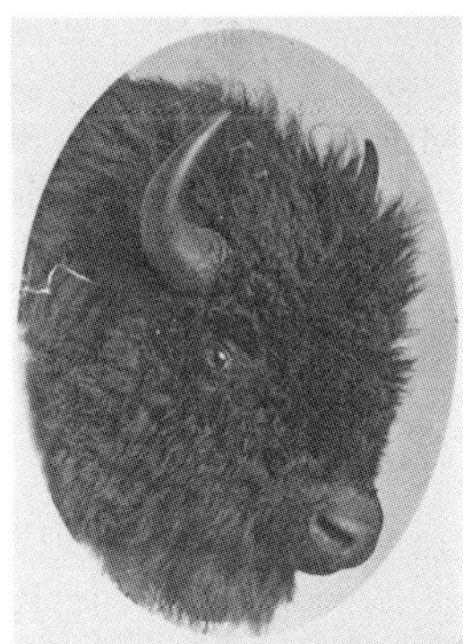

FIGURE 1.4B *William Henry Jackson,* Bull, *Albumen print, 1875, The Benton Museum of Art at Pomona College.*

legacies such as washing photographic plates in boiling hot springs and developing prints in portable darkrooms to create stealthy images of the American wilderness, crops the buffalo's head, which suggests its status as a dead trophy. In contrast, Muybridge utilizes and improves upon photographic technique to animate a dying species. Muybridge's animated cinematic stills can be likened to nucleic acids which, according to Lippit, transport information from one location to another and preserve the detailed function of every organ of creatures, which can be reanimated if configured in the right sequence (2001: 116).

Microcinematography

It is difficult to separate the history of DNA discoveries from advancements in photography, given that Rosalind Franklin's photographs using x-ray diffraction were pivotal in shaping the conception of the double helix (Mukherjee 2017: 153–4). Hannah Landecker inquires how the coupling of film, a time-based medium, with microscopes transforms the experience of scientists (2011). She observes that time-lapsed images enabled humans to perceive the "cellular temporal world" and the relations between cells and their environments, in contrast to stagnant images (2011: 381). Thus, cinema—and the instruments of microcinematography—have been crucial to cellular knowledge. The first time-lapse microscope was created in the 1890s by Étienne-Jules Marey, a French physiologist. Marey pleaded with Muybridge in a letter written to the magazine *La Nature* (1878) to help him capture birds in flight. He suggested that Muybridge animate his photographic strips and display them on screens, which would create a revolution (Ball 2013). In fact, time-lapse photography employs a similar method as Muybridge utilized to capture the horse's movement—the camera takes a photo at fixed intervals and strings them along in a sequence. Microcinemotography was perfected by Jean Comandon in Paris, who took thirty-two cinemagraphs per second of microscopic blood droplets, bacilli, and animals, enlarged and projected them, causing a flea to appear as large as a six-story house ("Microbes Caught in Action" 1909). Church Lab's Muybridge experiment is thus a continuation of the integral place of cinema in the production of knowledge of microbiology.

2. Do De-extinction Efforts, in the Context of this Essay, Impose Human Dominance or Restore Nonhuman Life?

Whereas time-lapse photography can animate microscopic entities, de-extinction experiments can reanimate life. Mirroring the premise of another film, *Jurassic Park*, the project Revive and Restore, inspired and supported by Church Lab, aims to use CRISPR to bring back from extinction ancient mammals, plants, and microbes (Mezrich 2017). In Church's quest to revive the woolly mammoth, its property of cold tolerance is characterized by placement of the letters A, G, G (Mezrich 2017: 53). Church argues that if one were to insert such combinations within the DNA of an ancestor such as the Asian elephant, one could create a woolly mammoth-like creature.

Although on temporary hiatus due to the Russia-Ukraine War, Church has been working in collaboration with Sergey and Nikita Zimov, the creators of Pleistocene Park in Siberia, north of the Arctic Circle, to restore the woolly mammoth to prevent carbon- and methane-filled permafrost from melting and thus causing ecological disaster (Bernstein 2022). Considered to be a "multispecies collaboration" by anthropologist Anya Bernstein (2022), the proposition is that humans alone cannot prevent this process. "[H]owever, pasture ecosystems can . . . Animals in pastures, looking for food, excavate and trample all snow several times each season, causing it to condense and lose its heat-insulating abilities" (Mezrich 2017: 120). For the Zimovs, the solution to climate change and mass extinction lies in animal ecosystems not in humans. However, as implicated in *Jurassic Park*, human dominance is still very much at play in the construction and genetic manipulation of such ecosystems.

3. Could Biotechnology Revitalize the Humanities by Breathing New Life into Works of Art, Literature, and Film? In What Ways Might a Humanistic Approach to this Experiment and CRISPR Allow Us to Think About Ethics?

Although interdisciplinary CRISPR research is sparse, scholars such as cultural anthropologist Eben Kirksey and literary scholar Everett Hamner have elucidated its connections to cultural products from Mary Shelley's *Frankenstein* (1818) to the *X-Men* franchise. Kirksey stresses the importance of ethics in the aftermath of Jiankui He's 2018 announcement of his rogue misuse of CRISPR to edit human embryos "Lulu" and "Nana," born in Shenzhen, China, through an in vitro fertilization pregnancy. The scientist claimed to have used CRISPR to disable a gene responsible for facilitating the entry of HIV into healthy cells, making them HIV-resistant. He violated several international scientific and ethical norms. He's clinical trial did not report prior studies, cite risks, follow guidelines in China or his own university, seek acceptable informed consent, or prove necessary as HIV is preventable in other ways (Krimsky 2019). Kirksey argues, "As scientists speculate about post-racial futures and nightmare military scenarios, as market forces bring new genetic technologies into the clinic at a dizzying speed, it is time to slow down and establish some clear rules for the road" (2021). Although this chapter has focused on CRISPR in animals and bacteria, and not humans, it is through an interdisciplinary approach that we may stand outside these issues and evaluate them thoughtfully and ethically, especially, as *Nope* (2022) highlights in the context of people of color, the dangers and responsibilities of human intervention.

In addition to the questions raised in this chapter, the Muybridge experiment questions the role that bacteria may play in digital media. Could mutations that occur in live organisms make the bacteria co-producers in this film's rerelease? What role do microbes play in this intersection of CRISPR, cinema studies, and animal studies? How do bacteria, viruses, and DNA subvert and reshape concepts such as memory and creativity as well as divisions between organic/inorganic, human/animal, and even the category of species?

Animal ethics, suffering, and environmental catastrophes involving CRISPR must also be considered. Emphasizing that our current era is one of unprecedented human control over life, Doudna writes: "Within a few decades, we might well have genetically engineered pigs that can serve as human organ donors—but we could also have woolly mammoths, winged lizards and unicorns. No, I am not kidding" (Doudna and Sternberg 2016). One must also consider the suffering that hybrid animals or animals brought back from extinction may experience. In this sense, we must affirm, as film scholar Anat Pick does, that vulnerability is a basis of radical ethics. As bioethicist Hank Greely discusses, this consideration means caring for creatures born or modified in labs, rather than abandoning them (2018).

CRISPR has been the topic of science-fiction media, such as the film *Rampage* (Brad Peyton 2018), starring Dwayne Johnson as a primatologist whose friend, an albino silverback gorilla, becomes transformed through CRISPR into a destructive beast. *Okja* (Bong Joon-ho 2017), is the fictional tale of an animal genetically engineered to increase meat production. The connection between CRISPR and cinema can help us imagine and mitigate nightmarish scenarios while experiencing empathy for those involved. Similarly, according to Doudna, "The humanities are incredibly valuable because writers have been thinking about these issues for a long time" (2018).

While CRISPR may help treat deadly disease and may fend off environmental extinction, the advantages and the risks of this technology currently cannot be separated from cinematic spectacle. De-extinction efforts are meant to combat shocking statistics such as the World Wildlife Federation's report that Earth has lost over half of its wildlife in the last forty years (Carington 2014). At the same time, one could argue that the Muybridge experiment is an imposition of human dominance into the very DNA of bacteria.

References

Adam, H.-C., ed. (2010). *Eadweard Muybridge, the Human and Animal Locomotion Photographs*. Taschen.

Ball, E. (2013). *The Inventor and the Tycoon: The Murderer Eadweard Muybridge, the Entrepreneur Leland Stanford, and the Birth of Moving Pictures*. Anchor Books

Bernstein, A (2022). "Pleistocene Park: Engineering Wilderness in a More-Than-Human World." Pomona College. Claremont: November 8.

Bliss, C. (2020). "Conceptualizing Race in the Genomic Age." In *For "All of Us"? On the Weight of Genomic Knowledge*, ed. J. M. Reynolds and E. Parens, special report, *Hastings Center Report* 50(3): S15–S22. doi: 10.1002/hast.1151

Brookman, P., and Braun, M. (2010). *Eadweard Muybridge*. Tate Publishing.

Carrington, D. (2014). "Earth Has Lost Half of Its Wildlife in the Past 40 Years, Says WWF." *The Guardian*, September 30. https://www.theguardian.com/environment/2014/sep/29/earth-lost-50-wildlife-in-40-years-wwf

Chow, A. R., and L. Z. (2022). "Nope, Explained: The Meaning of Jordan Peele's Latest Mystery." *Time*. https://time.com/6199449/nope-explained-meaning-jordan-peele/

Davis, J., and Shipman, S. (2016). "Chrononucleotides: Time Traveling with Dna | Joe Davis & Seth Shipman |Tedxbeaconstreet." *YouTube*, December 21. https://www.youtube.com/watch?v=VNezf_lCdi8

Doudna, J. (2018). "Gene Editing and the Future of Frankenstein. "*Frankenstein" Then and Now.* The Huntington. San Marino: 11 May.

Doudna, J. A., and Sternberg, S. H. (2018). *A Crack in Creation: Gene Editing and the Unthinkable Power to Control Evolution*. Mariner Books.

Greely, H. (2018). "Frankenstein and Modern Bioscience: Which Story Should We Heed?" *Modern Prometheus: Frankenstein in Film, Fiction and Contemporary Science*. Pomona College. Claremont: 27 October.

Hensley, M. (2017). "Race, Racing & Eadweard Muybridge's Photographs." Middleburg. https://www.middleburglife.com/race-racing-eadweard-muybridges-photographs/

Herbert, S. (n.d.). "Projecting the Living Image" Tate. https://www.tate.org.uk/context-comment/audio/projecting-living-image.

Jurassic Park (1993). Spielberg, S. USA: Amblin Entertainment.

Kirksey, E. (2021) *Mutant Project: Inside the Global Race to Genetically Modify Humans*. Black Inc.

Kolbert, E. (2015). *The Sixth Extinction: An Unnatural History*. Bloomsbury.

Krimsky S. (2019). "Ten ways in which He Jiankui violated ethics." Nature Biotechnology. January 3. 37(1): 19–20.

Landecker, H. (2011) "Creeping, Drinking, Dying: The Cinematic Portal and the Microscopic World of the Twentieth-Century Cell." *Science in Context* 24(3): 381–416. https://doi.org/10.1017/s0269889711000160

Lippit, A. (2008). *Electric Animal: Toward a Rhetoric of Wildlife*. University of Minnesota Press.

Mann, J. (2018). "The Photographer Who Gave Us the Moving Image." *Artsy*, January 10. https://www.artsy.net/article/artsy-editorial-eadweard-muybridge-moving-image

McDonald, S. N. (2022). "You've Seen 'Nope,' Now Let's Make Sense of It." *Andscape*. https://andscape.com/features/youve-seen-nope-now-lets-make-sense-of-it/

McHugh, S. (2011). *Animal Stories: Narrating Across Species Lines*. University of Minnesota Press.

Mezrich, B. (2017). *Woolly: The True Story of the Quest to Revive One of History's Most Iconic Extinct Creatures*. Atria Books.

"Microbes Caught in Action.; Moving Pictures of Them a Great Aid in Medical Research" (1909) *The New York Times*, October 31. https://www.nytimes.com/1909/10/31/archives/microbes-caught-in-action-moving-pictures-of-them-a-great-aid-in.html

Mukherjee, S. (2017). *The Gene: An Intimate History*. Vintage.

Nelson, A. (2016). *The Social Life of DNA: Race Reparations and Reconciliation after the Genome*. Beacon Press.

Nettleton, C. (2019). *The Artist as Animal in Nineteenth-Century French Literature*. Palgrave Macmillan.

Nope (2022). Peele, J. USA: Universal Pictures.

Peele, J., and Palmer, K. (2022). "GQ Hype: Jordan Peele and Keke Palmer." *GQ*. https://www.gq.com/story/gq-hype-jordan-peele-and-keke-palmer

Prodger, P. (2003). *Time Stands Still: Muybridge and the Instantaneous Photography Movement*. Oxford University Press.

Sample, I. (2017). "Harvard scientists pioneer storage of video inside DNA." *The Guardian*, July 12. https://www.theguardian.com/science/2017/jul/12/scientists-pioneer-a-new-revolution-in-biology-by-embeding-film-on-dna

Seyfried, G. (2008). "Mutants from Innerspace." *PolyCinease*. https://www.polycinease.com/mutants-from-innerspace/

Shimamura, A. P. (2002). "Muybridge in Motion: Travels in Art, Psychology and Neurology." *History of Photography* 26(4): 341–50. https://doi.org/10.1080/03087298.2002.10443307

Shipman, S. L., Nivala, J., Macklis, J. D., and Church, G. M. (2017). "CRISPR-Cas Encoding of a Digital Movie into the Genomes of a Population of Living Bacteria." *Nature* 547(7663): 345–9. https://doi.org/10.1038/nature23017

Zimmer, C. (2021). "A New Company with a Wild Mission: Bring Back the Woolly Mammoth." *The New York Times*, September 14. https://www.nytimes.com/2021/09/13/science/colossal-woolly-mammoth-DNA.html

2

Autopoiesis in Contemporary Bioart

Rethinking Autonomy and Agency

Charissa N. Terranova

If the relationship between art and autonomy was a central dialogue within twentieth-century criticism, then let that of art, life, and autonomy be a guiding force of that of the twenty-first century. Recognizing that art engages organisms, environments, ecologies, all kindred energies of life, as part of the discourse about art's necessary autonomy is not a contradiction in terms. The living within art does not negate the concept of autonomy because biological life is fundamentally *autopoietic*, or self-organizing, self-creating, and self-directing. Biological life *is* autonomous. It follows that if within twentieth-century paradigms of art criticism, autonomous art made autonomous life, then today autonomous life makes autonomous art.

For the philosophical purists, bringing living organisms to bear on art's autonomy carries the discussion deeper into the work of Immanuel Kant, that lodestar of thinking in terms of autonomous art, insomuch as the idea of an autopoietic, or self-organizing, organism emerges from Kant's writings. According to Andreas Weber and Francisco J. Varela, "Kant himself introduced the term 'self-organization' in its modern sense to biological theory" (1992: 99).

Kant identified the "self-propagating formative power" (*fortpflanzende bildende Kraft*) of living organisms to set in relief the nonmechanical, nonmetaphysical, and nonexternal organizational force of life itself (Molina 2010: 24). Life is thus not exactly a matter of "teleology without telos," or adaptation without design, to invoke the Scottish zoologist D'Arcy Wentworth Thompson (1992: 6), but rather a matter of design that is emergent from within the organism itself. Life is inherently creative in its autonomy. Unlike works of autonomous art that are the result of an artist acting upon materials, living autopoietic beings are *cause and effect of themselves*, and thus bear the autonomy of autopoiesis upon the art of which they are a part (Khurana 2013: 165).

Indeed, the various strains of contemporary bioart in this essay foreground the greater realm of the living and the richness of its situatedness within ecosystems, and by connection the threats of the Anthropocene. But the real crux of my argument sits elsewhere, within the rejection of various forms of scientific reductionism and anthropocentric art practices that emerge from the thesis that life is autopoietic. I argue, based on the 1973 writing about autopoiesis and the organization of the living by Chilean biologists Humberto R. Maturana and Francisco J. Varela, artists deploying microorganisms in their work fortify a narrative of life as interconnection, symbiosis, and evolution in real time. This is distinct from life defined by procreation, fitness, competition, survival, and evolution over deep time. Bioartists set in relief a story of life *other* than that of conventional Darwinism, natural selection, and its attendant reductionisms of the twentieth century, namely the selfish gene and gene triumphalism. At the center of this creative movement is autonomous autopoietic life itself, life that is alternatively self-interested and altruistic, but always generative and creative. It is also the life given shape in Charles Darwin's writings about beauty, sexual selection, and aesthetic agency, though this is not a topic I cover here.

In what follows, I define the term "autopoiesis," inscribing it within works of art that are dialectically autonomous and a matter of systems. They are autonomous by way of autopoietic life while at the same time systems-oriented insomuch as life itself is a system embedded in and imbricated with myriad systems. Thus, autonomous autopoietic life creates autonomous art through its systemicity. Updating critic Jack Burnham's writing about conceptual art and systems aesthetics in the late 1960s, I argue that this work constitutes a "living systems aesthetics" and examples of "real time living systems." With respect to the latter, I explore how artists work with bacterial mutations—that is, evolution unfolding in real time by way of the transformations of microbial life. By invoking Burnham's systems theory-based aesthetics, I set in relief the dialectical nature of life within bioart: how it is both autonomous in its critique of Darwinian reductionism while also held captive to systems of biotechnological reification within labs.

Autopoiesis and Systems Counterculture

Literary theorist Bruce Clarke identifies this social milieu as "systems counterculture" (2020: 101), embodied in popular culture by the first Earth

Day on April 22, 1970. The research-based expertise of systems counterculture coalesced within Stuart Brand's *Whole Earth Catalog*, published intermittently from 1968 to 1998 (2020: 101). While he never published in *Whole Earth Catalog*, systems theorist and art critic Jack Burnham was also a member of the systems counterculturati. While systems imply collectives, autopoiesis within biology focuses on the individual organism. Yet, the organism is what Varela would by the 1990s describe as a self without selfhood, or a "meshwork of selfless selves" (1991: 79–107).

The specifically anti-reductionist nature of Maturana and Varela's thesis coalesces perhaps counterintuitively around the individual not the system. This is already evident to readers in British cybernetician Stafford Beer's Preface to the text. "The subordination of the individual to the species cannot be supported," Beer argued (1980: 69). A new biology of complexity, other than genetics, had been born in the work of the two Chilean biologists when they argued that "biology cannot be used any more to justify the dispensability of the individual for the benefit of the species, society, or mankind, under the pretense that its role is to perpetuate them" (1980: 69). The new biology operated beyond the mechanism-vitalism dyad of the late nineteenth and early twentieth centuries. It also countered the late-twentieth-century mechanistic tendency to reduce life to genetic drives by locating a third way forward for the field. They hewed this path by defining the living organism—from the cell itself to single-cellular and multi-cellular organisms—as an autopoietic individual. Let it be clear that Maturana and Varela emphasized the individual not as a reinstatement of the rational individual of neoliberal economics, *homo economicus*, but rather to rescue the figure from the oblivion of Darwinism.

Maturana and Varela introduced the concept of autopoiesis in the early 1970s, in a cultural context open to experimental thinking about the living. The fate of the individual organism is not so much unknown, but rather camouflaged as the organism becomes part of larger data sets within population studies that reveal genetic drift over generations within evolution. "The species evolves," Maturana and Varela elucidate, "while the individuals are transient components whose organization is subordinated to its historical phenomenology" (quoted in Beer 1980: 69). Maturana and Varela clearly recognized how scientific paradigm shifts mold language, ideas, and practices of a given moment. The individual inscribed by science becomes the individual of society writ large. Maturana and Varela argued that Darwinism had created a "biological (scientific) justification" for "a society based on economic discrimination, competitive ideas of power and subordination of the citizen to the state" (1980: 118). The organism's autopoietic creative ability to equilibrate and self-maintain goes unmarked. For Maturana and Varela, this ability endows the individual organism with an identity—an agency—based on its intricate workings. "The organization of the individual is autopoietic," they declared, "and upon this fact rests all of its significance: it becomes defined through its existing, and its existing is autopoietic" (1980: 118).

Even while moving beyond the mechanism-vitalism debate that haunted biology throughout the twentieth century, the Chilean biologists adhered to the materialism connected to mechanism, describing the organism as an "autopoietic machine" as though to distance themselves from any sense of vitalism. Life at its most minimal is a metabolizing homeostatic cell—an autopoietic or self-creating biological machine rather than an allopoietic or other-created technological machine. This is life at its radix, *a priori* procreation and evolution. Maturana and Varela explain, using the language of holism, that as a "network of processes of production," autopoietic machines produce self-regenerating components that create and constitute the machine's existence within their interrelated network (1980: 79). Cellular life is at once open and closed, a combination of operational closure and environmental openness. "Even while autopoietic systems are environmentally open to material-energetic fluxes or semiotic mediations," Clarke explains, "their operations are internally closed so that the system sequesters its integrity as a functional unity" (2020: 39).

Living Systems Aesthetics

Such cellular life abounds within us and around us, but it evades visualization by the bare human eye. Giving visual shape and context to microbial life could be argued to be the *raison d'être* of bioart. Upon the invocation of bioart, many think of Petri dishes depicting a living form of the Mona Lisa having been carefully created by the microbiologist-cum-artist's guiding hand. Art historian-cum-artist Mariana Perez Bobadilla argues for a more contextual visualization of microbial life by returning to the use of the Windogradsky column for display of such minute lifeforms. A staple of public science education since their invention in the 1880s, scientists no longer use these seemingly architectural elements even while once used in laboratories (Dworkin 2020).

As long pillar-like stalks with wide colorful bands of striation indicating the diverse levels of ecological interchange, they are striking examples of the visual mediation of microbes in situ that offer an example of autopoietic life at the microscale given material form for perception at the mesoscale by humans. For Bobadilla and her collaborator Rodrigo Guzman Serrano, they offer a sense of the life of microbes that is closer to their own reality than when given material visual form for human use. "In comparison to the Petri dish," Bobadilla and Serrano explain, "the Winogradsky column provides a setting in which microorganisms can perform closer to the way they would do 'in the wild' under specific environmental conditions" (2020: 39).

Using the Winogradsky column as a means of working with microbes presents life as fundamentally autonomous but embedded and interconnected. Here, microbes are presented less in terms of instrumentality and transaction and more according to "multi-species interpretation" (2020: 39). The

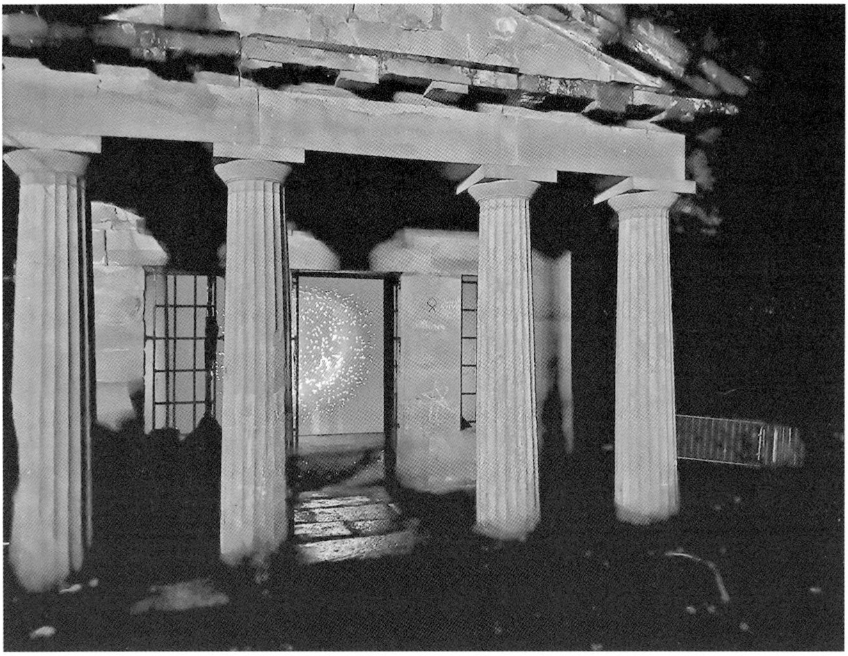

FIGURE 2.1 *Louise Mackenzie,* The Stars Beneath Our Feet, *2015. Site-sensitive audio-visual installation, 7:07 (loop). Commissioned for Lumiere Durham 2015, a four-day international light festival produced by Artichoke in the UK. Image courtesy of the artist.*

microbes living in Winogradsky columns are both autonomous and embedded: they are self-organizing systems living within and adjacent to other self-organizing systems.

The same is true of the microorganisms at work in the UK-based artist Louise Mackenzie's art practice (see Chapter 7 in this volume). Mackenzie's site-sensitive audiovisual installation of 2015, *The Stars Beneath Our Feet* (Figure 2.1) sonifies and visualizes microorganisms for human experience. Working with scientists at Durham and Northumbria Universities, Mackenzie used light-absorbing microalgae, ancient photosynthetic organisms that absorb light and convert it into oxygen, to create a cosmic experience within an eighteenth-century Georgian folly. Using advanced microscopy techniques, Mackenzie, Richard Thompson, and Paul Vickers captured as data the disturbances created by microalgae and translated this information into an immersive audio-visual experience. "These microscopic organisms," Mackenzie explains, "helped to create the conditions that enabled life on earth and have the uncanny quality of resembling a sky filled with stars when viewed at scale" (2015).

The autonomous autopoietic microalgae operate in this context in the biotechnological systems of university laboratories. They are living systems

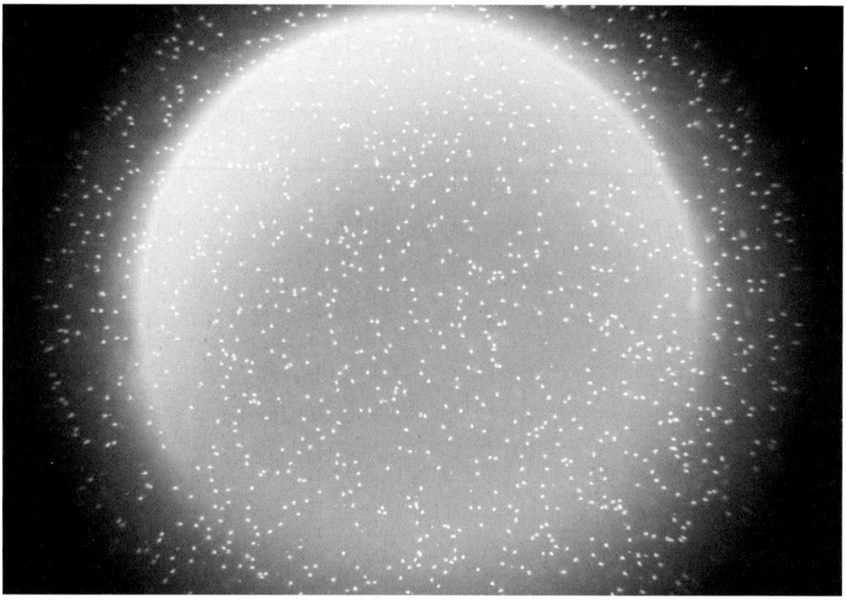

FIGURE 2.2 *Louise Mackenzie,* The Stars Beneath Our Feet, *2015. Video still. © Louise Mackenzie, 2015.*

working within artificial systems. Mackenzie's use of microorganisms to create art, an activity that is both collaboration and transaction, constitutes a living systems aesthetics. In the September 1968 essay, "Systems Esthetics," Jack Burnham identified the then new tendency for conceptual artists, such as Joseph Kosuth, Les Levine, and Robert Morris, to adumbrate systems rather than craft objects. These artists abjured the conventions of making hallowed objects, opting instead to make work in a variety of systems-based media, including, for example, a billboard, close-circuit monitors, and the landscape itself. No longer a matter of "modern formalism" and its "craft fetishism," this was work that participated in the world of information flows and real time ecologies, human and otherwise. He located connections between the hands-off de-skilling at work, for example, in Moholy-Nagy's "telephone paintings," Kaprow's evanescent happening, the General Systems Theory of Austrian theoretical biologist Ludwig von Bertalanffy, and the latest technologies of the Rand Corporation and DARPA to identify a new mode of making art that was more about ideas, attitudes, technology, and critique rooted in language games. Burnham's systems aesthetics were part of the zeitgeist. In the same years, art critics Lucy Lippard and John Chandler identified this tendency as the "dematerialization of art" (1968).

Based on the theoretical biology of von Bertalanffy, Austrian author Arthur Koestler coined the neologisms of "holon" and "holarchy" to identify autonomous, self-organizing systemic units—holons—hierarchically organized within other greater system units—holarchies (1970, 131–54).

Real Time Living Systems: Bioart and Evolution in Real Time

Burnham expanded his exploration of systems aesthetics into the temporal, publishing the essay "Real Time Systems" exactly one year later in September 1969. For Burnham, art had become virtually indistinguishable from the ecological, citing the "real time work of art" of the bloody 1969 Berkeley *People's Park Protest*, where thousands of people marched and brought sod to turn an abandoned lot into a park. Burnham (1969) asks: "In a country of 100,000,000 vehicles, what better gallery could you find?" Burnham's examples of art, by connection, were conceptual, alive, and cybernetic, ranging across mediums and scale, from Hans Haacke's "Chickens Hatching" to Dennis Oppenheim's "September Wheat Project" to Les Levine's short-lived Manhattan restaurant. So too is the autopoietic life in examples of contemporary bioart by Ken Rinaldo and Anna Dumitriu (see Chapters 8 and 19 in this volume). This work, distinct from the conceptual art of Burnham's moment, configures another real time system, namely evolution itself, a topic to which I return below.

American bioartist and Director of the Art and Technology Program at Ohio State University Ken Rinaldo's *Enteric Consciousness* (2012) is an interactive work, the centerpiece of which is an orange tongue-shaped massage chair (see Chapter 8 in this volume). The work plays on the seminal role played by the flora in the gut in the expression of human behavior—from consciousness itself to genetic function to mood swings. Adjacent to the chair sits a large glass-shaped stomach, which is filled with the living bacteria *Lactobacillus acidophilus*. If the microbial cultures are living, then one can enjoy a 15-minute massage in the chair. Otherwise, the chair is dead still.

In the 2011 project *Communicating Bacteria*, British bioartist Anna Dumitriu explores the many communicative behaviors of microbes. With this work, she combined bioart, historical textile techniques, and 3-d mapped video projections to show how different types of bacterial communication, including "quorum sensing (voting on issues affecting the colony and signaling their presence to other bacteria), chemotactic signaling (detecting harmful or favorable substances in the environment), and plasmid exchange (e.g., for transfer of antibiotic resistance genes)," could be expressed for human audiences to see and understand (2021). One finds evidence of bacterial communication in textile designs stained with dyes made from bacteria that change color depending on their behavior, crochet patterns based on bacterial responses, and a series of hacked antique whitework embroidered pieces created using genetically modified bacteria. This body of work by Dumitriu and her collaborators plays out the behavioral possibilities of monocellular autopoietic life, in particular, as it elides into multicellular existence by quorum sensing, which, as Dumitriu explains, enables the group to express specific genes only at particular population densities. Underscoring the dialectic of autopoietic life and systems Dumitriu tells her audience: "processes controlled by quorum sensing are ones that are unproductive when undertaken by an individual bacterium but become effective when undertaken by the group" (2021).

Elsewhere in her practice, Dumitriu interacts with bacterial diversification and antibiotic resistance. The project *Made Do and Mend* (2017) is an installation by the artist, consisting of a vintage Second World War suit, related era framed pamphlets urging austerity and recycling, and her mother's toy sewing machine. The suit has embroidered patches of material made splotchy through the artist's use of bacterial lifeforms and the new CRISPR-Cas9 molecular gene-mutation technology. The installation piece pays homage to the discovery of penicillin by the Scottish doctor Alexander Fleming in 1928, its use during the Second World War, and then its potential obsolescence by way of antibiotic-resistant strains of bacteria like *Escherichia coli*. Implicit in these bioart practices is further investigations of autopoietic life at work in real time evolution—that is, art work and material culture illustrating the life of bacteria, which reproduces itself through horizontal gene transfer happening moment by moment in actual unfolding time, in contrast to the vertical gene transfer of larger organisms such as mammals, the mutations of which are evidenced across population studies in phylogenetic time.

Conclusion: Biotechnological Systems and Paul Vanouse's "Labor"

Unique to contemporary bioart, as opposed to conceptual art practices of the twentieth century, is its foundational relationship to systems of biotechnology. The autonomy of autopoietic life is fundamentally connected to this relationship insomuch as without biotechnology, bioartists would have little to no access to living microorganisms. While there would still be bioart without microorganisms, there is no bioart without biotechnology or biohacking. The autonomy of autopoietic life within bioart emerges perhaps counterintuitively from its proximity to contemporary biotechnologies.

While built upon it infrastructurally, bioart's critical-intellectual relationship to biotechnology is nuanced, even ambiguous. As challenging and often esoteric, works of bioart take time to engage with and fully understand. They often require a viewer to do research, both on materials and techniques since doing work in a laboratory is a large part of the meaning-making at work in a given work of bioart. The opacity of bioart bestows it with a slowness that goes directly against the grain of biotechnology, the development of which is rapid-fire and the use of which, while painstaking in terms of skill and knowhow, is similarly intended to bring quick results. As this work is part of what I've elsewhere described as

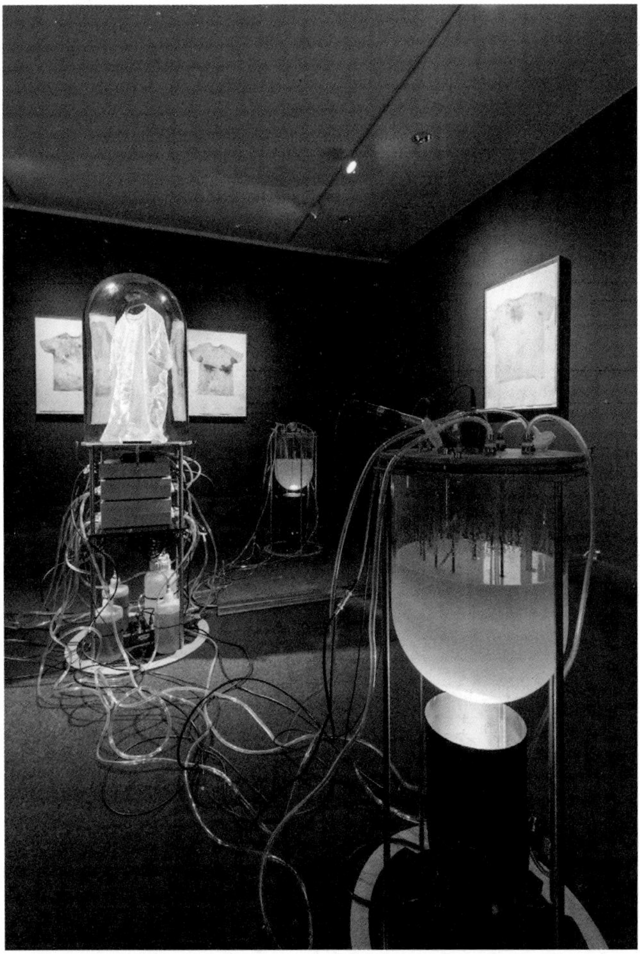

FIGURE 2.3 *Paul Vanouse,* Labor, *Multi-sensory installation, Prix Ars Electronica Festival, OK Center, Linz, Austria, 2019. Installation shot: CYBERARTS OÖ Kulturquartier, Linz. Photo by Otto Saxinger. Image courtesy of the artist.*

"nonstop modernism," bioart is an example of what Lutz Koepnick calls "slow modernism" (2014: 15). Even while bioart is in part constituted by the latest inventions in biotechnology, it is nonetheless at the same time a critique of its accelerationism. Because bioartists use such biotechnology for purposes other than profit further places them at odds with the element of scientific reductionism tied to profit-based results. As such, bioart is often, simply put, useless to scientists. As a contemporary avant-garde—a trifecta of neo-dadaism, neo-conceptualism, and newfound design-based problem solving—bioart's "use" is bound up with its importance to poetic thinkers and cultural critics and its actual utility in terms of bench science understood as a set of design skills. An intentional goal of many bioartists is popular science education—teaching people the facts of science while providing subtle critical analyses of neoliberal biotechnological and biopharmaceutical systems and entities.

American bioartist and Program Head of Emerging Practices in the Art Department of SUNY Buffalo Paul Vanouse's *Labor* (2019) (see Chapter 17 in this volume) bears autopoietic life that brings home its autonomy at three valences: in its posthuman relativizing of life; in how it pushes against the accelerationism of scientific reductionism and biotechnology by its very use of scientific reductionism and biotechnology; and in its powerful presence as a polysensual immersive installation work. Through the use of three bioreactors, the work *Labor*, according to Vanouse, "endeavors to re-create the scent of human exertion" (2020: 32).

The three bioreactors incubate "a species of human skin bacteria responsible for the primary scent of sweating bodies: *Staphylococcus epidermis*, *Corynebacterium xerosis*, and *Propionbacterium avidum*" (2020: 32–3). While human sweat itself is odorless, these bacteria that nourish themselves on it, "create volatile, odiferous chemical compounds," which humans connect to the smell of working bodies. The work bears its physicality in varied dimensionalities and the olfactory and optical senses alike. As a work that is explicitly about physical human labor, it comments on the political economy of human labor in the twenty-first century, and its ever-decreasing utility in the face of ever-increasing automation, both at the scale of the globe. Vanouse reframes human identity as hybrid and symbiotic, revealing that humans are not simply hosting other organisms but collaborating and developing a cooperative relationship with them. The classical *cogito ergo sum* gives over to a new perspective: *I am because I am of microorganisms*. Their evolution is our evolution.

This work tells of a world where human sweat is on a path to desuetude not unlike human physical labor itself. Here, the smell of human sweat must be manufactured by bioreactors in an art gallery instead of in the body. But it is not to say that human labor is obsolete, for its obsolescence is not biologically preordained but rather a matter of human regulation and choice. The autonomy of autopoietic life within bioart reinforces its systems-based relationship to the public sphere and its responsibility, both human-to-human and organism-to-organism.

One must look to history to understand the greater value of bioart, which shares a genealogical space with the nineteenth-century sciences of biology, geology, and ecology in that romanticism is its forebear. Romanticism proposes an ontological position alternative to enlightenment thinking, one by which "humanity depends upon and is part of nature rather than standing outside it" (Stone 2014: 42).

Contemporary romantics seek to de-accelerate capitalism to recognize the earthen-ness of human life. The de-accelerationist motto is: humans do not stand above nature: humans are nature. Through the lens of history, three critical transformations are evident. First, bioart is part of a posthumanist, nature-centric paradigm shift in the humanities, sharing space with current academic studies such as animal studies, plant studies, and the medical humanities. As with the medical humanities, bioart elucidates the workings of highly complex science and its pedagogical functions. Second, within art history, bioart braids Dadaist critical irony and the cerebralism of conceptual art to lay bare the utility and corruptions at work in biotechnology. Bioart shifts from art into design, gaining the potential of environmental remediation and other types of amelioration. Third, bioart operates beyond the history of art within the realm of the lived now of function. In the end, just as Marshall McLuhan and György Kepes thought, artists will not be made redundant by technologies like artificial intelligence. Instead, they become leaders. She is shaman of old and new worlds like.

References

Beer, S. (1980). Preface to "Autopoiesis: The Organization of the Living." In *Autopoiesis and Cognition: The Realization of the Living*. D. Reidel Publishing, 69.

Bobadilla, M. P., and Serrano, R. G. (2020). "Microorganisms on Stage: Winogradsky Columns as Performative Displays in Art and Science." *Performance Research* 25(3): 39.

Burnham, J. (1969). "Real Time Systems" *Art Forum*.

Clarke, B. (2020). *Gaian Systems: Lynn Margulis, Neocybernetics, and the End of the Anthropocene*. University of Minnesota Press, 101.

Dumitriu, A. (2021). "Communicating Bacteria," unpaginated text provided to author by artist on January 3.

Dworkin, M. (2020). "Sergei Winogradsky: A Founder of Modern Microbiology and the First Microbial Ecologist." *FEMS Microbiology Reviews* 36(2): 36479.

Khurana, T. (2013). "Life and Autonomy: Forms of Self-Determination in Kant and Hegel." In *The Freedom of Life: Hegelian Perspectives*. August Verlag, 155–93.

Koepnick, L. (2014). *On Slowness: Toward an Aesthetic of the Contemporary*. Columbia University Press, 15–52.

Koestler, A. (1970). "Beyond Atomism and Holism—the Concept of the Holon." *Perspectives in Biology and Medicine* 3(2): 131–54.

Lippard, L., and Chandler, J. (1968). "The Dematerialization of Art." *Art International* 12(2): 31–6.

Mackenzie, L. (2015). Press Release. https://www.loumackenzie.com/the-stars-beneath-our-feet

Molina, E. (2010). "Kant and the Concept of Life." *CR: The New Centennial Review* 10(3): 21–36.

Roqué, A. J. (1985). "Self-Organization: Kant's Concept of Teleology and Modern Chemistry." *The Review of Metaphysics* 39(1): 107–35.

Stone, A. (2014). "Alienation from Nature and Early German Romanticism." *EthicalTheory and Moral Practice* 17(1): 42.

Thompson, D. W. (1942). *On Growth and Form*. Cambridge University Press, 6.

Weber, A., and Varela, F. J. (1992). "Life after Kant: Natural Purposes and the Autopoietic Foundations of Biological Individuality." *Phenomenology and the Cognitive Sciences* 1: 99.

Vanouse, P. (2020). "Labor: The Post-Anthropocentric Body 'At Work'." In *Performance Research*, ed. J. Hauser and L. Strecker, 25(3): 32.

Varela, F. (1991). "Organism: A Meshwork of Selfless Selves." In *Organism and the Origins of Self*, ed. A. I. Tauber. Kluwer Academic Publishers, 79–107.

3

The Exterminating Angels

Bio-/Thanatos-Art

Pablo Baler

Life does not speak, it listens and waits.

A THOUSAND PLATEAUS, GILLES DELEUZE AND FELIX GUATTARI, 1980

In Luis Buñuel's 1962 Mexican surrealist film, *The Exterminating Angel*, a group of aristocrats become mysteriously trapped in a mansion and, day by day, regress to the most brutal state of life: what starts as a posh cocktail gathering turns into violence, suicide, death, sacrifices, hallucinations—total anarchy.

In 2017, Swedish director Ruben Östlund's movie, *The Square*, takes on a similar theme of "conventions versus instincts" but adds a twenty-first-century twist. The movie, a satirical take on the contemporary art world is, more to the point, an exploration of the division between human and nonhuman. Interestingly, in this film, not only does a human appear as a gorilla, but all boundaries seem stretched and threatened to the point of collapse: a chimpanzee is presented as a child living with a woman, a baby is treated as a pet accessory, a dog plays a detective, and the director of the museum assumes the role of a pig throughout the movie.

At any rate, both surrealist films in their own way can be seen as commentaries on the pervasive impulse to categorize life, an impulse that undergirds our ethics, our politics, our science, even our metaphysics. In fact, it is this impulse to categorize, divide, and classify that inspires one of the ever-present questions in bioart scholarship: What does it mean to be human?

This question, invariably asked and answered by humans, has become a trite, preemptive fuzzword that has served to reveal our own limitations at attempting a radical rethinking and deconstruction of the concept of *being human*.

I bring up these two films because the question of the boundaries between human and nonhuman—as thematized in both of them—does point to a central issue touching both the implications of CRISPR-Cas9 and art. In terms of CRISPR—and synthetic biology, in general—one could ask at what point this rewriting/editing/proofreading/repairing of our genome will threaten our understanding of what it means to be human. Through my examination of contemporary bioart practices by Oron Catts and Ionat Zurr, Stelarc, and Eduardo Kac, coupled with writings of philosophers Giorgio Agamben, Michel Foucault, and Gilles Deleuze, I argue that at the same time, this underlying tension evokes an issue that is central to aesthetics: the conflict between conventions and subversions, between law and desire, between grammar and poetry.

This radical rethinking and deconstruction of the very concept of *being human* is also at the heart of Agamben's biopolitical perspective. Agamben traces back these categorizations that have produced the concept of "human" to the original assumptions underlying the historical foundations and formations of the category of "life." This is the focus of Agamben's interpretation of biopolitics, which departs from that of Foucault. Agamben is not interested so much in the ways bodies and populations are subjugated and regulated by specific, politicized institutions, but more broadly on how the conceptualization of life itself has had, from its inception, a political dimension. In his book, *The Open: Man and Animal*, Agamben writes: "[E]verything happens as if, in our culture, life were what cannot be defined, yet, precisely for this reason, must be ceaselessly articulated and divided" (2004: 13).

According to Agamben, the divisions between life and nonlife, vegetative and relational, organic and animal, and the essential one between animal and human, create a mesh of oppositions that can be interpreted as the very conditions of possibility of Western culture and the inherent mechanisms underlying power. Agamben presents our contemporary models of power as a reflection of this original gesture, the foundational threshold of our culture: negating our animality. Agamben writes:

> Man exists historically only in this tension; he can be human only to the degree that he transcends and transforms the *anthropophorous* animal which supports him, and only because, through the action of negation, he is capable of mastering and, eventually, destroying his own animality. (2004: 12)

Thus, Agamben prepares the ground for a fundamental question about the relationship between culture and power:

> What is man, if he is always the place—and, at the same time, the result—of ceaseless divisions and caesurae? It is more urgent to work on these

> divisions, to ask in what way—within man—has man been separated from non-man, and the animal from the human, than it is to take positions on the great issues, on so-called human rights and values. (16)

This broader perspective on biopolitics might cast a deeper and wider net of connections than Foucault's biopolitics. In addition, exploring what Agamben calls "human values"—that is, deconstructing ethics—would not only shed light onto that mysterious boundary that separates man from animal but also unsettle the "anthropological machine," in Agamben's words, that makes the very question of "what does it mean to be human" possible.

Some experiences in the field of bioart seem to merge (even if not always, and not always intentionally) with that objective of deconstructing the ethical dimension underpinning any categorization/conceptualization of life. In the works I will consider here, this exploration manifests itself as resistance against categorization and conceptualization. The bizarre, surprising, poetic, sometimes repulsive nature of some of these artistic experimentations has increasingly turned bioart into a stage where the tensions between Power-Knowledge and Resistance are not only played out but also spectacularized.

For Foucault, the idea of "power-knowledge" is central to this question. We can arrive at a better understanding of this term in Eugene Thacker's essay, "Open Source DNA and Bioinformatic Bodies." Here, Thacker considers biotechnology through its artifacts: genome databases and sequencing computers, gene discovery software, DNA chips, and online data mining. Thus, he defines biotechnology as a set of practices that strategically recontextualizes biological components and processes. In turn, this recontextualization defines our current understanding of "life" in ways that are artificial, codified, reductionist, and instrumental (governed by optimization and standardization). Beyond mere philosophical or pure scientific curiosity, Thacker foregrounds the ability of biotechnology to target specific genes for diagnostics, to develop gene-based drug therapy and pursue more genetic research. These—diagnostics, therapy, and research—Thacker points out as *three central areas of economic [and power] concentration in the biotech industry*:

> The point is to understand the thread . . . that connects ontological statements about the bio-informatic body with the regulation and control of genetic data; the connections between ontological statements about what a body can do (it can be encoded, databased) and the political attributes of those bodies in relation to notions of property (our clones, our selves), value (medical and/or economic value), and power (biology as technology). (2007: 35)

To take CRISPR-Cas9 as one instance of technological "property, value, and power," we already see how it is playing out (also spectacularly) in

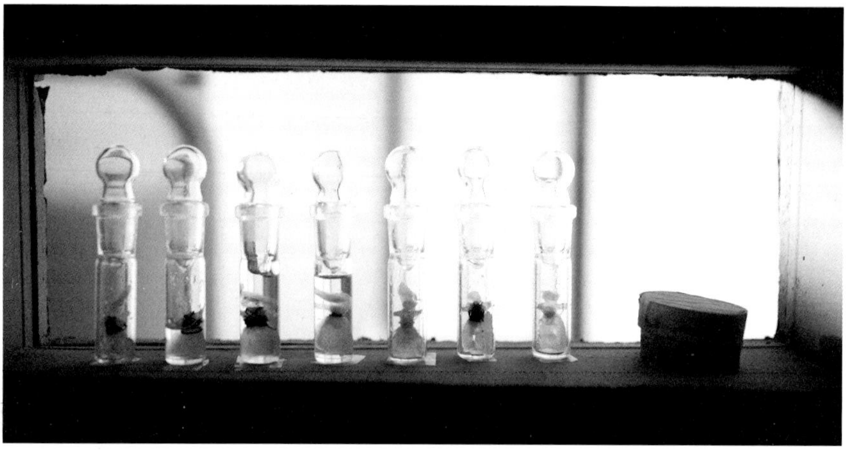

FIGURE 3.1 *Oron Catts and Ionat Zurr (The Tissue Culture & Art Project),* The Semi-Living Worry Dolls, *Cells, Biodegradable/bioabsorbable polymers and surgical sutures, 2000. Courtesy of The Tissue Culture & Art Project (Oron Catts & Ionat Zurr).*

terms of legal battles and the uncertainty moving forward about ownership of copyright and license applications for medicine, agriculture, and industry.[1] In the final analysis, whoever wins this battle will eventually control the technology and reap the unfathomable financial and political benefits that this particular technology will yield.

It is against the background of these connections between the ontological dimension of bodies and life, and the control (technological, informational control) of bodies and life, or in more general terms, against the background of a foundational conflict between "free play and becoming" (Ziarek 2004) versus power and optimization that we should revisit some of the works done by currently practicing bioartists.

Bioart

The work by Oron Catts and Ionat Zurr could be seen as one of the most relevant in terms of the exploration of the ethical dimension underpinning the conceptualization of life. This team of wet-biology artists, based in West Australia, have also grown "semi-living" sculptures, by culturing cells around degradable polymer scaffolds, as in the *Semi-Living Worry Dolls* project (Catts, Zurr and Ben-Ary 2000) inspired by Guatemalan worry dolls. They have been exploring this concept of semi-living entities since 2000. About the impact of this particular project, they wrote in *Leonardo*, back in 2002:

> The ethical questions that have been raised by the project mainly concern our relationships with these semi-living objects: Are we going to care for

> them? Do these entities contribute to the objectification of living organisms? Their existence calls into question long-held belief systems and our perceptions of life and death . . . [O]ur goal is to create a vision of a future where some objects are partly artificially constructed and partly grown/born in order to generate a debate about the directions in which biotech can take us. (2002: 366)

Among other ideas that were experimented with in the *Semi-Living Worry Dolls* exhibit was to have members of the audience kill these semi-living beings, thus prompting them to commit a form of "semi-cide" for which we still have no legal (or even ethical) categories to conceptualize.

In this context of the ethical exploration at the intersection of the arts and the biological sciences, bioartists Jason Knight and Jennifer Willet, founders of the Canadian art collective BIOTEKNICA, together with Ionat Zurr and Oron Catts, worked on the project titled *TERATOLOGICAL PROTOTYPES* (2006/7). For this project, they worked with a cell line taken from a teratoma in order to create what Jason Knight described as a "three-dimensional tissue, living sculptural specimen." A teratoma is an unusual and grotesque form of cancerous growth where—similar to an embryo—cells differentiate and divide producing a living blob that can contain multiple tissues like hair, skin, bone, vascular systems, and more complex organs or processes such as torso, hands, feet, teeth, even a rudimentary, beating heart.

Etymologically, teratoma means a tumoral monstrosity; yet what is truly monstrous about it is not the growth itself but the fact that it defies identity and conceptualization. The teratoma is both a symbol and a manifestation of openness. As philosopher Max Marcuzzi has said, "Teratology is biology in its wildest form."[2] Possibly inspired by this work, Stelarc, who conceives his own aesthetics in terms of *alternative anatomies and architectures*, wrote:

> Only when we can bio-print or stem cell grow, caressing a lump of teratoma-like living tissue that is slimy, that is throbbing, that quivers to the touch will we have a potent object to interrogate what it is to be alive, what it means to be a body and what it means to be human. (2013: 133–4)

The term itself, "bioart," or better yet "sci-art," which is the general label under which bioart is usually presented (science + art), already reveals a tension between the instrumental, reductionist, and rational apparatuses of science coupled with the allegedly counterproductive, chaotic, and transformational forces of art. This tension, if not completely manifested, is hinted at in the art projects presented above, as they unsettle the instrumental, regulatory, rationalizing thrust of power-knowledge as well as the "anthropological machine" that undergirds bio-techno-politics.

If we see art as an alternative form of knowledge, as a speculative epistemology, then we could conceive of it as an opening, a crack in the politics of knowledge from where thinking otherwise suddenly becomes possible. Now, the question of this tension, at the intersection of science and art, between no-power and art practice (which has been central to the historical avant-garde), is once again unavoidable: Has art become an extension, another manifestation, of the techno-scientific organization of modernity (biopower and information technology) or is there still in art the potential for a radical disruption of the technicity of power?

In *The Force of Art*, Krzysztof Ziarek rephrases this question in the context of bioart by asking "to what an extent transgenic art is complicit with the manipulative flows of power or whether, on the contrary, it exposes, complicates, or perhaps even contests them" (2004: 96). Ziarek is referring here to the work of Eduardo Kac, but this is also a relevant question as CRISPR-Cas9 will be slowly introduced to the palette of expressive resources used by a new generation of bioartists.

Ziarek's answer to the question about the conflict between complicity to and resistance against the technicity of power is mindful of the complex and ambiguous mechanisms at play. On the one hand, it shows a necessary connection between art and *the flows of power*, and, on the other, it foregrounds the transformational energy of art, and in particular bioart. Referring to the work of artist Eduardo Kac, he writes:

> In a way, the power is still the same—it is the power to transfer genes and engineer transgenic life forms—and yet its momentum appears to be different: geared no longer just to manipulation, that is, to further intensification of the reach of power into the microelements of being, but rather to the possibility of a different, "artistic" disposition of forces. (2004: 98)

This "artistic" disposition of forces transforms the instrumental modality of techno-power into the "unproductive" modality of "letting be": "Art's force is not an alternative power but an alternative to power, which releases forces into the element of reciprocal free play and becoming" (2004: 12).

Ziarek's understanding of the "forces of art" not as an alternative power but as an alternative to power is a useful framework to think about bioart as a deconstructive force. However, in relation to the "technological deployment of power that is regulative of modern life," Ziarek makes only a distinction of degree and not of principle between the critical and transformational forces of the historical avant-garde and twenty-first-century bioart. According to Ziarek, both the historical avant-garde and twenty-first-century bioart bring about a critical tension between art and technology by disclosing "an alternative to the paradigms of production, mobilization, and technical manipulation at the core of contemporary operations of power" (2004: 3).

I would contend, however, that in order to articulate the potential impact of bioart (in contrast to the historical avant-garde) we would have to take into account a historical shift that is not considered in Ziarek's interpretation. We need to distinguish between the type of metaphorical "alternative to power" operating in the historical avant-gardes[3] and the type of enacted "alternative to power" that is intrinsic to biological art as it interferes with life itself and with the concrete anxieties produced by this interference. Taking into account that bioart is not merely "representational" but also *in vivo*, the creations of bioart are not merely metaphorical but impinge upon the fabric of life itself. This paradigmatic shift from representation to instantiation also requires rethinking the conventional divide between art and reality and therefore reframing the question of the nature of resistance in art. For in this context, ethics emerge as an unexplored boundary where the dialectics of resistance, as the condition of possibility of art, is suddenly reconfigured.

Some bioart projects, such as Oron Catts and Ionat Zurr's semi-killing of semi-living beings, Stelarc's alternative anatomies, Eduardo Kac's chimeric trans-species, or Jason Knight and Jennifer Willet's teratological prototypes, already carry some seeds of enacted resistance and somehow fulfill bioart's potential as an alternative to power not only to the extent that they threaten our assumptions and our categorizations and conceptualization of life—and the biopolitically charged dichotomy of man and animal—but more fundamentally to the extent that they bring to life the unconceptualizable, the uncategorizable, the unthinkable, the unimaginable.

Applied ethics and bioethical enquiry perpetuate traditional, utilitarian, humanistic principles when answering new questions brought about by biotechnology. For instance, what type of rights should we give to clones purely grown for organ harvesting purposes? To what extent should we genetically modify species as a method to fight epidemic outbreaks—especially in light of the COVID-19 pandemic? In regards to a technology like CRISPR, that allows us to edit genomes with the same ease we do a Word document, should we establish a standard grammar, a style manual, a genetic rhetoric we should all follow when toying around with specimens and species? What are the style guidelines in terms of how much we can modify/enhance/upgrade ourselves for optimization purposes? Bioart offers not only unsettling answers to these questions but it unsettles the principles of reductionism, essentialism, instrumentalism, and optimization preserved by biopower as a way to explain away the dangers inherent in even settling these questions.

In an article in *E-Flux*, Boris Groys points to this thrust of subversion in contemporary art. He writes:

> Traditionally, we associate art with a movement towards perfection. The artist is supposed to be creative. And to be creative means, of course, to bring into the world not only something new, but also something better—better functioning, better looking, more attractive. All these expectations

> make sense—but . . . in today's world, all of them are related to design and not to art. Modern and contemporary art wants to make things not better but worse—and not relatively worse but radically worse: to make dysfunctional things out of functional things, to betray expectations, *to reveal the invisible presence of death where we tend to see only life*. (2014, online; emphasis is mine)

Taken to its unstable extreme, artistic expression should invalidate the society in which it occurs and, therefore, it should invalidate itself. Thus, it is coextensive to its nature for bioart to display its ethical imagination in terms of this dialectics of creation and self-destruction. In bioart, which is also a Thanatos-art, this impulse *to reveal the presence of death* where we tend to see only life should be understood as yet another way of threatening the ethics of life management and optimization.

By stretching to the breaking point, the boundaries of the biological imaginary and the hierarchical system of life that underwrites biopolitics, bioart has the potential to create new relational meanings and transcend the merely metaphoric gesture of avant-garde resistance. As opposed to the utopian biopower of universal knowledge, absolute diagnostics, quantum computing, limitless therapy, eternal reversal of aging, total de-pollution, resurrection of species, and infinite energy production, these artists (whether they know it or not) defy optimization and traditional ethics in favor of dysfunctionality and disruption. Accordingly, one could fathom bioart as not only involving the killing of semi-living beings (as in Ionat Zurr and Oron Catts' work), the proliferation of cancerous monsters, the arbitrary reconfiguration of dysfunctional anatomies (Stelarc), or the growth of chimeric life forms (Kac), but also taking poetic licenses in, for example, the production of oil-spilling bacteria, the dissemination of designer-viruses, microorganisms genetically targeted to give the final blow to species in danger of extinction, or conjuring up entire new ecologies with their own evolutionary, hierarchical, and relational logics.

It is from the perspective offered by these dysfunctional and dystopian projects that we could revisit the question of "what does it mean to be human." But to revisit it not to simply articulate an answer but to disarticulate the question. In contrast to the "anthropological machine" that underwrites biopolitics, bioart (or one could call it "biofiction") could develop in terms of a "zoological speculation machine" that could commit to offering paradoxical, contestable, contradictory ways of thinking about that recurring question.

Biofiction

In his advocacy for excess and singularity beyond generalization and difference, Foucault also proposes, in *Theatrum Philosophicum*, a vindication of uncommon sense, of ill will that resonates with this alternative ethics of disruption, or, to put it in George Bataille's terms, with this aesthetics of evil. Foucault writes:

> [C]ommon sense extracts the generality of an object while it simultaneously establishes the universality of the knowing subject through a pact of goodwill. But what if we gave free rein to ill will? What if thought freed itself from common sense and decided to function only in its extreme singularity? What if it made malign use of the skew of the paradox, instead of complacently accepting its citizenship in the *doxa*? (1970)

As long as bioart will not make "malign use of the skew of the paradox," it will remain a didactic reflection, a form of political activism (as in Steven Kurtz and the Critical Art Ensemble), a commentary (as in Eduardo Kac's provocative pieces), a metaphorical critique of current possibilities in biotechnology (as in Paul Vanouse's *Operational Fictions*). As long as bioart remains an extension of modern technicity, it will not fulfill its inherent disruptive potential as an instantiated form of resistance against biopolitics (both in its narrow Foucaultian and in its more general Agambenian sense). This resistance will manifest itself not only in stretching the boundaries of bodies and life forms but more radically in stretching the boundaries of our ethical imagination.

These posthuman, dystopian aesthetics of instantiation have the potential to bring us closer to Deleuze's vision—mentioned in the epigraph—of life as not speaking . . . as listening and waiting. This is the tension I've been trying to articulate. Deleuze and Guattari's complete sentence reads: "Language is not life; it gives life orders. Life does not speak; it listens and waits" (1987: 76). Here language is understood as an example of the grammar and syntax of domination[4] while listening and waiting are manifestations of the openness to the infinite Other outside and inside, before and after, to the Other that looms in the always partial process of becoming. It is from here that we can start thinking about replacing not just the definitions of "human" but even the question of "what is human" with something else, but replacing it not with another question but with a suspenseful listening and waiting which is the only suitable response in the face of "becoming" that is life—and death—itself.

Notes

1 The last decade has witnessed a legal battle between a team at Berkeley, led by Jennifer Doudna and Emmanuelle Charpentier, and another at the Broad Institute of MIT and Harvard University, led by Feng Zheng, regarding royalties for CRISPR-Cas9. The United States Patent and Trademark office has awarded in favor of the Broad team on multiple occasions. Doudna and

Charpentier, winners of the Nobel Prize in Chemistry for their research on CRISPR-Cas9 in 2020, uphold the rights in Europe as of 2021. Legal issues involving CRISPR are likely to become more complex as potential therapies for human use continue to develop (Ledford 2022).

2 Quoted in Thacker, *Signs of Life*, p. 63.

3 As one instance one could cite Kandinski's abstraction as a response against what he saw as the breakdown of the world by discoveries in the nature of the atom; or the shift away from materialism into occultism.

4 Or as Donna Haraway (1991) puts it in "The Cyborg Manifesto," "grammar is politics by other means."

References

Agamben, G. (2004). *The Open: Man and Animal*. Stanford University Press.

Catts, O., and Zurr, I. (2002). "Growing Semi-Living Sculptures: The Tissue Culture & Art Project." *Leonardo* 35(4): 365–70.

Deleuze, G., and Guattari, F. (1987). *A Thousand Plateaus: Capitalism and Schizophrenia*. University of Minnesota Press.

Foucault, M. (1970). "Theatrum Philosophicum." *Critique* 282: 885–908. https://www.generation-online.org/p/fpfoucault5.htm

Groys, B. (2014). "On Art Activism." *e-flux journal* #56. http://www.e-flux.com/journal/56/60343/on-art-activism/

Haraway, D. J. (1991). *Simians, Cyborgs, and Women: The Reinvention of Nature*. Routledge.

Ledford, H. (2022). "Major CRISPR Patent Decision Won't End Tangled Dispute." *Nature* 603(7901): 373–4. https://doi.org/10.1038/d41586-022-00629-y

Stelarc (2013). "ALIVENESS & AFFECT: ALTERNATE ART & ANATOMIES." In *The Next Thing: Art in the Twenty-First Century*, ed. P. Baler. FDUPress.

Thacker, E. (2007). "Open Source DNA and Bioinformatic Bodies." In *Signs of Life: Bio Art and Beyond*, ed. E. Kac. MIT Press.

Ziarek, K. (2004). *The Force of Art*. Stanford University Press.

4

Viral Variation(s)

Juan Eduardo Cirlot and the Poetics of Permutation

Paul Cahill

Reading and Revising the "Book of Life"

Conceiving of "[t]he human genome . . . as an information system and, more specifically, as a 'Book of Life' written in the language of DNA, or DNA code, to be read and edited," as scholars have done since the middle of the twentieth century, has had a significant impact both in and beyond the life sciences (Kay 2000: 1). This essay will explore one potential example of this phenomenon by tracing how the experimental work of Spanish poet Juan Eduardo Cirlot (1916–73) might be read in the context of DNA and gene editing. In *The Poetics of DNA*, Judith Roof frames DNA as a specific "information system," namely a textual one in which "DNA is imagined as something that stands for something else in the same way that words or images refer to concepts or objects by representing them" (2007: 73). Such an approach is possible thanks to important work by scientists like physicist George Gamow, who asserted in a piece in *Nature* (1954) that "the hereditary properties of any given organism could be characterized by a long number written in a four-digital system" and that "enzymes . . . are long peptide chains formed by about twenty different kinds of amino-acids, and can be considered as long 'words' based on a 20-letter alphabet" (318). For Gamow, "the question arises about the way in which four-digital numbers can be translated into such 'words'" (318). A key consequence of understanding the genetic code in linguistic terms was the possibility of editing it as text, since "[b]eyond the academy, as captured in the pages of the *New York Times* and

Time magazine, the genetic code also signaled the potential for genetic engineering, even before the advent of recombinant-DNA technologies in the 1970s" (Kay 2000: 3).

Periodicals circulating in Spain during the 1950s, 60s, and 70s also discussed these topics. In May 1969, in *La Nueva España*, Diego Luis Hortelano published a three-story installment about DNA, the possibilities of genetic engineering, and its potential for the future of humanity. The overarching title of the series, "La ciencia más allá de la ficción" (Science Beyond Fiction), sets the tone for the series. The second installment begins by tracing the history of scientific discoveries regarding the concept and structure of DNA, framing it in linguistic terms (Hortelano 1969: 17). Similar predictions regarding the possibilities of modifying this text appeared in the Spanish press of the 1970s, including the following passage from a piece that appeared in the popular magazine *Blanco y Negro* in 1971: "Starting from this moment, infinite possibilities open themselves up to genetic science. Who can doubt that in the future mankind may be able to program his own cells? Mankind will have at its disposal means to realize important modifications in the human being" (Carrión 1971: 58).[1]

The awarding of the 2020 Nobel Prize in Chemistry to Emmanuelle Charpentier and Jennifer Doudna recognized the quantum leap in this field that the development of CRISPR represented. These developments—seeing DNA as a text and gene editing—have an artistic parallel in the work of artists like Joe Davis (1950–), Heather Dewey-Hagborg (1982–), Eduardo Kac (1962–), Joel Ong (1983–), and Ken Rinaldo (1958–). Spanish artists and writers also use their work to both reflect on scientific developments and explore new possibilities. A precursor in this field was Salvador Dalí, whose painting entitled *Galacidalacidesoxiribunucleicacid (Homage to Crick and Watson)* (1963) engaged directly with the work of Francis Crick and James D. Watson.

Spanish Poets and Science

Agustín Fernández Mallo (1967–), an experimental physicist and an accomplished writer, has explored the relationship between poetry and the sciences in his poems and essays. His *Postpoesía. Hacia un nuevo paradigma* (2009) (Postpoetry. Toward a New Paradigm), which offers a brief history of engagements with the sciences by poets and other artists before elaborating a theory of "postpoetry," proposes a posthumanist poetic practice that is open to paradigms, ideas, and ways of thinking associated with other fields. According to Fernández Mallo, "the relationship between science and written poetry is there and has always been there, although as a parasitic rarity of *correct* poetry" (2009: 30).[2] His own poetic practice explores these same questions, most notably in the case of *Antibiótico* (2012) (Antibiotic). Fernández Mallo's work is possible in part because of the work of another poet, Clara Janés (1940–), whose poetry dialogues with several different fields, including mysticism, philosophy, mathematics, and physics. A recent study of Janés's poetry explores its "determined combination of mystical and scientific discourse," exploring how this poet's work "links these disparate ways of knowing by explicitly incorporating scientific thinkers and their ideas into her mystical poetry and by highlighting scientific language as a historically embedded discourse" (Faszer-McMahon 2021: 142). A key precursor, and the focus of this chapter, is the groundbreaking work of Spanish poet Juan Eduardo Cirlot.

A significant figure in Spanish literary and intellectual life during the middle of the twentieth century, Cirlot is perhaps best known beyond Spain's borders for his *Dictionary of Symbols*, originally published in Spanish in 1958. As a poet, Cirlot is largely a cult figure, both because of the unconventional character of his work and the fact that much of his poetry was either self-published or published in small batches. Since the turn of the twenty-first century, readers have had greater access to Cirlot's work thanks to the publication of *Bronwyn* in 2001, which brought together his entire "Bronwyn" Cycle, and the subsequent publication of the two volumes *En la llama. Poesía (1943–1959)* (2005) and *Del no mundo. Poesía (1961–1973)* (2008), followed by an anthology by Spanish poet and editor Elena Medel: *El peor de los dragones. Antología poética, 1943–1973* (2016). Critical engagements with Cirlot's poetry have situated it in the contexts of its ludic potential (Debicki 1994: 79–80), its mystical connections (Janés 1996), its relationship to music (specifically 12-tone music) (Gracia and Ródenas 2010: 414–15; Manjón and Schmitt 2006: 529), its connection to political resistance to the Franco regime (Ledesma 2016: 147–8), its dialogue with numerology (Fernández and García 1999: 172), and its connection to painting and other visual arts (Lorenzo 1993: 193–4). A context that has yet to be explored by critics, though, is the potential dialogue between Cirlot's poetry and scientific discourses and developments taking place during the 1950s, 60s, and beyond.

One year after George Gamow's piece on DNA and protein structures appeared in *Nature*, in 1955 Cirlot explored the potential of recombining a set of elements in the form of a poetry collection that uses its first poem as a starting point for a series of variations and permutations. While the first two permutations in *El palacio de plata* (The Silver Palace) only entail the reordering of verses, the nine variations that follow break up these verses but still use the same words as the original text (2005: 511–25). Cirlot's later work would go even further, dividing words into letters and sounds. The culmination of this trajectory is his *Variaciones fonovisuales* (Phonovisual Variations), written in 1972 and published in 1996. This collection is made up of a series of texts that use a name or other word as a starting point to create a series of "phonovisual" variations of this initial word.

These poems use the open space of the page as a medium to explore the multiple axes and dimensions of letters, sounds, and words, and how they interact with one another. His phonovisual variations explore the ways in which different letters and types of letters operate (vowels vs. consonants), explore the impact that different sequences of letters have, and challenge the reader with respect to how to divide, segment, and "read" different sequences of letters. Beyond such challenges, however, his poems also examine the ways in which letters and sequences mutate and change. My readings of a series of poems from *Variaciones fonovisuales* seek to merge a thematic and formal engagement with scientific discourse, showing how the techniques they employ engage with some of the same questions at stake in both contemporaneous and future scientific developments, regardless of whether this poetry was created with these developments in mind.

What Cirlot refers to as a "fonetista," or phoneticist style, focuses on individual letters and minimum units of sound and meaning rather than words, verses, or sentences. These permutations usually employ found texts in the form of names as a starting point for a series of variations and permutations and in some cases introduce an additional set of letters. The two principal names are "Bronwyn," a character played by actress Rosemary Forsyth (1943–) in Franklin Shaffner's film *The War Lord* (1965) and "Inger," after Swedish-American actress Inger Stevens (1934–70). The permutations of these names take two primary forms, which could be said to parallel to a degree the ways in which genetic mutations take place. The first is a more mechanical and systematic permutation in which names are resequenced according to a fixed rule, while the second employs a more aleatory principle. Excellent examples of both forms can be found in *Inger, permutaciones* (1971) (Inger, Permutations). As Cirlot's prologue to this collection explains, "[t]he first part of *Inger, permutaciones* is made up of the 120 permutations that the name Inger provides (1 × 2 × 3 × 4 × 5). The second part is made up of free combinations formed with the phonic material that comes from these permutations" (2008: 659).[3]

Serial and Procedural Forms, or The Diverse Ways in which Items Come Together

Forms like these remind us of the "serial" and "procedural" forms Joseph M. Conte discusses in *Unending Design: The Forms of Postmodern Poetry* (1991). Paradigm shifts effected by scientific developments play a key role in his framing of procedural and serial poetry, since "[t]he concept of serial form," he argues, "could not be convincing without the 'granular' physics of subatomic particles and molecular combination and recombination" (1991: 19). "The open form of the poetic series," for Conte, "is defined by its limitless set of relations; it takes its shape from the diverse ways in which items come together, undetermined by (in Umberto Eco's terms) an 'external *necessity*' that would prescribe its organization, or by the 'internal' necessity that is the claim of organic form" (15). "A procedural form," on the other hand, "is 'closed' by virtue of its entirely predetermined structure, but the function of that structure is radically different from that of traditional closed forms" (15). Rather than closing off avenues for exploration, "[t]he procedural form is a generative structure that constrains the poet to encounter and examine that which he or she does not immediately fathom, the uncertainties and incomprehensibilities of an expanding universe in which there can be no singular impositions" (16). This structure resists the "hierarchical cosmos in which all things can be known and situated accordingly" that Conte associates with more traditional forms, in the process echoing what "[t]he series as an open form—with its aleatory and indeterminate qualities" can do to challenge the traditional construct of "an organic sequence that still hopes to discover an immanent form and a unity in creation" (15–16). Advances in scientific accounts of the different ways in which genes mutate offer potential parallels to the serial and procedural poetic forms discussed by Conte. Gregor Mendel, for example, "understood that genetic crosses obey the rules of probability—the same rules that apply when tossing coins, rolling dice, or drawing cards" (Simon 2015: 187). Genetic mutations can be spontaneous and random (echoing serial forms), or the result of intentional manipulation (echoing procedural forms): "[b]ecause mRNA is read as a series of nucleotide triplets during translation, adding or subtracting nucleotides may alter the triplet grouping of the genetic message. All the nucleotides after the insertion or deletion will be regrouped into different codons" (219). I would argue that Cirlot's phonovisual variations incorporate elements from both forms discussed by Conte, allowing them to bridge the gap between form and content and present an early potential engagement with—and perhaps contribution to—the scientific discourses and questions upon which later writers and artists are able to build.[4]

Cirlot's *Phonovisual Variations*

The version of *Variaciones fonovisuales* published in 1996 is a facsimile of the typescript produced in 1972, one year after the appearance of *Inger, permutaciones*. Each variation, not numbered and occupying its own page, uses a name or other text as a starting point from which to create a "phonovisual" variation of this initial text. Cirlot's "phonovisual" variations use the open space of the page as a medium through which to establish a productive dialogue between written and verbal expression by exploring the multiple axes and dimensions of letters, sounds, and words, and how they interact with one another. The implicit questions these variations ask and explore include how different types of letters work with one another, whether individual letters can represent entire words, and if the sequence of letters and words matters. Taken together, these experiments create space for a potential dialogue between Cirlot's poetry and developments in the study of DNA and gene editing.

1. Eight Phonovisual Variations on the Name of Inger Stevens (1935–1970)

The first of the "Ocho variaciones fonovisuales sobre el nombre de Inger Stevens (1935–1970)" (Eight Phonovisual Variations on the Name of Inger Stevens [1935–1970]) opens with a distinction between two types of letters by presenting Inger Stevens's full name as the sum of consonants and vowels:

INGER STEVENS +
NGR STVNS + IEEE (Cirlot 1996: 17)

Subsequent variations in this same section explore tensions between the relative importance of each type of letter, with the fifth and seventh variations presenting either consonants or vowels as primary. Keeping the first and last name separate, the fifth variation presents two versions of Inger Stevens's full name without vowels, implicitly inquiring whether the sequence of both names and their consonants matters:

NGR
STVNS
SNVTS
RGN (1996: 21)

The seventh variation of the "Inger Stevens" sequence shifts from a focus on exploring the interaction between types of letters (vowels and consonants) to one centered on the interaction between Inger Stevens's first and last names. This variation ends with a figure made up of two columns that opens with an initial split between the letters from "Inger" and "Stevens," as well as an initial comingling of vowels and consonants. The right column mixes elements from "Inger" and "Stevens," beginning with an emphasis on consonants before producing rows comprised entirely of vowels:

IN	V S T E
GER	S V T E
	T V S E
IN	V T S E
REG	S T V E
	T E V E
IN	E E E E
GRE	E E E E
IN	I I I I
RGE	I I I I (1996: 23)

Both the fifth and seventh Inger Stevens variations employ combinations of different numbers of letters: three and four letters in the fifth; two, three, and four in the seventh. Both also explore different sequences of a delimited set of letters, including sets like: NGR, STVNS, IN, GER, and VSTE.

At the middle point of the figure—the sixth permutation in the right-hand column—a breakdown in this system occurs, which brings the four-letter sequence closer to the original text from which these letters were drawn—"Stevens"—by removing the *S* from the preceding line and adding an *E* between the *T* and *V*. In a sort of procedural effect, this greater proximity to the base term "Stevens" is followed by a sort of viral multiplication of vowels in the form of two lines comprised entirely of the letter *E*. This breakdown is in turn accompanied by another breakdown in the pattern presented up until this point. While the pattern had been the combination of "IN" and a variant of "GER" in the left column accompanied by a series of three sequences of *V*, *S*, *T*, and *E* in the right column, after the third such instance of this pattern—which consists entirely of the vowel *E*—there is no additional sequence in the right column. A closer look at this pattern reveals that every third sequence in the right column determines the first letter in the next sequence by moving the second letter to the first position in the next sequence:

TVSE ---> VTSE
STVE ---> TEVE

While the first example is a transposition of the first two letters, this is not the case in the second example, which reveals a mutation of sorts in which the *S* becomes an *E*.

Any attempt to trace or posit patterns in this figure runs into the sticking point of how to divide these sequences and decide what unit of measurement should be used. While it seems logical to divide the left column into units of two, because they maintain a clear division between letters belonging to the different syllables that make up the name "Inger," the column on the right presents a more complex situation. These sequences suggest the possibility of tracing patterns between two or three sequences, while simultaneously offering reasons to disqualify both options. While separating these sequences into pairs that echo the permutations of "Inger" seems logical, readers are left with the question of what to do with sequences that do not have one opposite them in the left column, as is the case with the third and sixth sequences. Dividing the sequences on the right into sets of three for the purpose of analysis would seem to be an interesting alternative, but even this method runs into trouble when we keep in mind that there are only two such sequences and that the third row from the bottom does not have a sequence in either column.

Things become even more complex when we consider the column on the right as four separate columns composed of different letters. This way of framing Cirlot's text reveals several interesting details. In the last column, for instance, we see that every single row has a vowel in this column (either *E* or *I*). If we move one column to the left, we notice a series of vertical pairs created by virtue of the place they occupy within each sequence (in this case the third letter in each four-letter sequence): TT, SS, VV, EE, and II. Since these pairs appear even in rows in which there is no "Inger" element, they seem to be independent of the anchoring presence that these pairs seem to exert on the figure. The eight variations that make up the first section of *Variaciones fonovisuales* thus explore both differentiation and duplication, demonstrating that these phenomena take place in both systematic and seemingly random ways. Despite the complexity of these variations, though, it is worth pointing out that the raw material they work with is the name "Inger Stevens."

2. Eight Phonovisual Variations on the Name of Helma with the Addition of 5 Letters (I,Y,D,O,S)

A quick glance at the title of the second section of *Variaciones fonovisuales* reveals a key difference between it and the sections that precede and follow it. Rather than limiting itself to the five letters that make up the name "Helma," this section introduces five additional ones: *I*, *Y*, *D*, *O*, and *S*. Although the technique of using individual letters to stand in for words or series of words is used in all three sections of *Phonovisual Variations*, the most straightforward iteration of this technique is found in the fifth variation on "Helma," where a series of five words is flanked by a single capital letter:

Hada	D
Ella	L
Lila	L
Maha	H
Ala	L (1996: 31)

This figure highlights the significance of the placement of letters in two different ways. From a purely visual standpoint, the sequence of *D*, *L*, *L*, *H*, and *L* stands out, and upon closer inspection we see that these same letters appear in the second-to-last position of the word from which they are taken. If we focus on the other end of the figure, we see that together the first letters of each word spell out the word "Helma."

While the act of creating space for readers to find "Helma" in this figure seems to fit clearly within the text and its project—producing variations on the name "Helma"—what seems less clear is what can be drawn from the letters that make up the column on the right. They are all consonants, and, except for the first one (*D*), all come from "Helma" rather than the set of letters added to "Helma" in this series of phonovisual variations (*I*, *Y*, *D*, *O*, *S*). The fact that *H* is one of the letters in this series stands out, since in multiple instances in this text, the letter *H* is presented as extraneous or at the very least not as significant as other consonants. The letter *L*, in contrast, is presented as central. When we look beyond the *H* and the *D* that belong to the set of "imported" letters, we notice that we are left with three *L*s: the third—and middle—letter in "Helma," whether we count from the beginning or end of the word.

If we keep in mind the somewhat hidden presence of the word "Helma" in the words in the column on the left as we examine the first variation, we will find another moment where the reader has to work to identify and fill in patterns. The end presents this base term in its full form followed by a series of four pairs of letters from this name:

HELMA
MA
LA
ME
LE (1996: 27)

With some work from readers, this otherwise abstract figure reveals a series of interesting details while simultaneously raising a series of questions. Perhaps the first detail that stands out is that each pair contains a consonant and a vowel, in the same order (consonant, vowel). A further look reveals a phenomenon like what we observed in the case of the seventh variation on "Inger Stevens," and specifically the right column which featured vertical pairs of repeated letters ("TT," "SS," "VV," "EE," "II"). In the first variation on "Helma," the pairs we find are "AA" and "EE." The letter that stands out because of its absence in these pairs is the "H" with which "Helma" begins, and a closer look at these pairs and the invisible work that generated them shows the superfluous role that "H" plays in this case. This hidden work can be summed up in the following schematic:

MA: (H) E L M A
LA: (H) E L M A
ME: (H) E L M A
LE: (H) E L M A

Multiple acts of mirroring take place in the generation of these pairs. If the first two lines create more distance between *A* and the consonants to its left, the opposite effect is at work in the final two lines. One key element that comes across in all four of these examples is the fact that "H" does not count as an element, since we move through a sequence of four letters instead of five. Permutations of the base word upon which this series is founded seem to work primarily by way of selection and subtraction, in the process highlighting the relative importance of some letters over others. The act of identifying patterns and the expression of the series's base term relies heavily on work done by readers. In a process not unlike the work of unraveling the secrets of the genetic code, the final series of *Variaciones fonovisuales* will ask readers to do just as much work—if not even more—as they end up administering a series of poetic and linguistic experiments of sorts.

3. Nine Phonovisual Variations on the Name of Bronwyn

The final series of variations included in *Variaciones fonovisuales* employs many of the same elements as the earlier ones did but turns its exploration of techniques into a test to see how far certain forms of expression can be pushed while still maintaining the same outcome. If several of the variations on "Helma" seem to be asking whether the "H" plays a role in the expression and production of sound and meaning, at least two of the variations on the name "Bronwyn" ask whether different letters can be used to produce the same sound—and meaning—as others do. Both the first and fourth of these variations explore whether the combination of the vowels *U* and *I* can be used in place of the combination of the consonants *W* and *Y* used in the traditional spelling of "Bronwyn."[5] While the fourth variation focuses on presenting alternatives by way of a visual juxtaposition,

WY WY WY WY
UI UI UI UI
. . .
WYN
UIN (1996: 40)

the first variation undertakes a deeper and more explicit juxtaposition and hierarchization of these two alternatives. This variation uses parentheses to establish an implicit hierarchy between these two versions: "BRONWYN (BRONUIN)" (1996: 37). The exploration of these alternatives accompanies the text's exploration of some of the other key questions addressed by these variations, namely whether order matters, and the categories into which letters can be divided. Thus, we see two versions of the implicit question regarding the roles played by vowels and consonants as well as an implicit question regarding the importance of order,

BRN BRN
OUI OWY

NYWNORB (NIUNORB) (1996: 37)

in the process framing the initial juxtaposition of "BRONWYN" and "(BRONUIN)" in the larger context of questions addressed by Cirlot in his Phonovisual Variations.

What may have appeared to be implicit questions up to this point are followed by an explicit one regarding a specific instance of the placement and sequencing of letters in an intriguing example. While the pair made up by "BRONWYN (BRONUIN)" and "NYWNORB (NIUNORB)" raises the question of reading sequences of letters—whether poetic or genetic—forward or backwards and whether these different sequences make a difference, the following example is joined up to the ones we have discussed so far and asks an entirely different question:

<table>
<tr><td>BRN
OUI</td><td>BRN
OWY</td><td></td><td></td></tr>
<tr><td colspan="2">NYWNORB (NIUNORB)</td><td>¿WRONBYN?
W___B
BRONWYN
B___W (1996: 37)</td><td>N

O</td></tr>
</table>

The material on the right side of the figure asks whether the positioning of certain letters matters for the expression and meaning of a word. The letters in this case are two consonants, "B" and "W," which also happen to be the first letters of the two syllables that make up the name "Bronwyn." The focus of this question—the positioning of "B" and "W"—is highlighted by the use of an underscore that stands in for the rest of the word, a placeholder which implicitly undermines the importance of the other letters in the word. In this final section of *Variaciones fonovisuales*, Cirlot takes full advantage of the potential that the visual arrangement of words and letters has to draw readers into the text vis-à-vis determining how language works and produces sound and meaning. While this engagement with the subtleties of written and spoken expression undoubtedly takes place within a poetic and linguistic context, this anchoring of the text certainly does not preclude other implications, especially when readers consider his series of variations as a whole.

Conclusion: Hereditary Heterodoxy

A significant potential implication of poetic experiments like Juan Eduardo Cirlot's *Variaciones fonovisuales* is the opening up of literary discourse to engagement(s) and dialogue with other fields of inquiry. Even if Cirlot's work does not engage directly with scientific discourse or developments, its exploration of the possibilities of permutation, repetition, and the introduction of outside elements into an existing system make a compelling case for identifying potential parallels between poetry and scientific advances in genetics and gene editing before, during, and after the time when Cirlot was writing. In light of recent developments in gene editing like CRISPR, work like Cirlot's illustrates how literature and the arts can serve as a way of imagining new possibilities in other fields: possibilities that may not exist yet but nonetheless inspire us to keep pushing the envelope in other fields. Just like "the 'granular' physics of subatomic particles and molecular combination and recombination" (Conte 1991: 19) played a key role in the formation of postmodern poetry, the experiments undertaken by Cirlot set the stage for later Spanish writers like Clara Janés and Agustín Fernández Mallo to engage in more direct explorations of scientific concepts by merging form and content and employing both serial and procedural forms and strategies.

Acknowledgment: I would like to thank Óscar Curieses for his generous gift of the copy of Juan Eduardo Cirlot's *Variaciones fonovisuales* that made this study possible.

Notes

1 "A partir de este momento se abren infinitas posibilidades a la ciencia genética. ¿Quién puede dudar de que en el futuro el hombre sea capaz de programar sus propias células? El hombre dispondrá de medios para realizar modificaciones importantes en el ser humano" (Carrión 1971: 58).

2 "la relación entre ciencia y poesía escrita está ahí y siempre ha estado, aunque como rareza parasitaria de la poesía *correcta*" (Fernández Mallo 2009: 30).

3 "[l]a parte I de *Inger, permutaciones* se compone de las 120 que da el nombre Inger (1 × 2 × 3 × 4 × 5). La parte II se compone de libres combinaciones formadas con el material fónico procedente de tales permutaciones" (Cirlot 2008: 659).

4 Important later works that explore the poetic potential of DNA include those by Joe Davis (the original DNA poet), Eduardo Kac's *Genesis*, and Louise Mackenzie's *Genophone*.

5 Playing with alternatives in this manner might be compared to the notion of redundancy in the genetic code, in which up to six different triplets of DNA base pairs can represent one amino acid (as discussed by Joe Davis in this volume).

References

Cirlot, J. E. (1996). *Variaciones fonovisuales*. Barcelona: La Central.

Carrión, I. (1971). "La genética en la nueva frontera de la medicina." *Blanco y Negro*, October 9, 58–61.

Cirlot, J. E. (2005). *En la llama: Poesía (1943–1959)*. Siruela.

Cirlot, J. E. (2008). *Del no mundo: Poesía (1961–1973)*. Siruela.
Conte, J. M. (1991). *Unending Design: The Forms of Postmodern Poetry*. Cornell University Press.
Debicki, A. P. (1994). *Spanish Poetry of the Twentieth Century: Modernity and Beyond*. University Press of Kentucky.
Faszer-McMahon, D. (2021). “Science in the Works of Clara Janés: A Poetics of Theoretical (Meta)physics.” In V. L. Ketz, D. Smith-Sherwood, and D. Faszer-McMahon (eds.), *A Laboratory of Her Own: Women and Science in Spanish Culture*. Vanderbilt University Press, 141–65.
Fernández, E., and García, E. (1999). “Cirlot: poesía permutatoria y abstracción.” *Cuadernos Hispanoamericanos* 589–590: 165–77.
Fernández Mallo, A. (2009). *Postpoesía. Hacia un nuevo paradigma*. Anagrama.
Fernández Mallo, A. (2012). *Antibiótico*. Visor.
Gamow, G. (1954). ‘Possible relation between deoxyribonucleic acid and protein structures.’ *Nature* 173(4398): 318. doi:10.1038/173318a0
Gracia, J., and Ródenas, D. (2010). *Historia de la literatura española*. Vol. 7. Crítica.
Hortelano, D. L. (1969). “El milagro de las ‘instrucciones imperfectas.’” *La Nueva España*, May 28, 17.
Janés, C. (1996). *Cirlot, el no mundo y la poesía imaginal*. Huerga y Fierro.
Kay, L. E. (2000). *Who Wrote the Book of Life?: A History of the Genetic Code*. Stanford University Press.
Ledesma, E. (2016). *Radical Poetry: Aesthetics, Politics, Technology, and the Ibero-American Avant-Gardes, 1900–2015*. State University of New York Press.
Lorenzo, A. J. (1993). “La poesía de Juan-Eduardo Cirlot a la luz del informalismo.” *Revista de Filología de la Universidad de La Laguna* 12: 191–9.
Manjón, D., and Schmitt, T. (2006). “‘Mi voz en el sonido de tu luz,’ Estructuras musicales en la poesía de Juan-Eduardo Cirlot.” *Bulletin of Spanish Studies* 83(4): 523–39.
Roof, J. (2007). *The Poetics of DNA*. University of Minnesota Press.
Simon, E. J. et al. (2015). *Campbell Essential Biology with Physiology*. Pearson.

PART TWO

Biotechnology and the Arts: Practice

5

Baitul Ma'mur

DNA Manifolds and the House of Angels

Joe Davis

How might we generate 2.417 hundred million billion angels on the head of a pin? The project *Baitul Ma'mur* traces the origins of, and our human neurological predisposition toward, religious belief, before proposing a poetic gesture for the COVID-19 pandemic that blends science and spirituality. *Baitul Ma'mur* was conceived as a means of honoring the millions of tragic global deaths due to COVID-19 in a manner that transcends racial or religious difference. In a multidisciplinary collaboration, including Pakistani artist and biologist Sarah Khan, the project integrates Muslim religious belief within scientific practice. Blending science with aesthetics, Khan and I employed the repeating structures found in Islamic art to define new, high-density forms of nested DNA information storage that we describe as *DNA Manifolds*. Within this nested structure we stored multiple repeats of the phrase "Subhan Allah" that, when written or uttered according to Arabic tradition, brings forth an angel. Thus, the *Baitul Ma'mur* project, through increasing the existing information storage capacity of DNA, has the potential to manifest billions of angels, as a gesture of comfort in the face of the devastating impact of COVID-19 on lives around the globe.

The Human Predisposition for Spirituality

Several years ago, on a visit to San Francisco, California, I had the good fortune to meet an unusual character who inspired me to pursue a project that would reach across languages and cultures and touch on science, art, religion, and mathematics. Yassir Chadly is a Sufi teacher and scholar. I've made it a point to visit Yassir whenever I've traveled to the region. Yassir fixes tea, and we proceed to while away hours in long discussions about "secrets of the universe" and broad topics of divinity—the kind of conversations I personally find fascinating.

In fact, the topic of divinity can assume more forms than most of us tend to realize. An intriguing case in point is that many Neanderthal grave sites, found in Europe and the near East, have been determined to be intentional. One of the most fascinating and controversial burial sites is the Shanidar cave in Iraqi Kurdistan.[1] The *Shanidar IV* remains were laid to rest sometime between late May to early July with at least eight species of flowers, mainly small brightly colored wildflowers. It is probably not coincidental that flower types present in the burial suggest that they may have been chosen for their respective medicinal properties.[2]

For context, we can note that in recent years, considerable attention has been devoted to relationships between neuroscience and human spirituality. Proponents of the neuroscience of religion say there is a neurological and evolutionary basis for subjective experiences traditionally categorized as spiritual or religious (McNamara 2009). Specific parts of the human brain seem to be associated with religious experience. In the 1990s, neuroscientists Leslie Ruttan, Stanley Koren, and Michael Persinger stimulated these parts of the brain with a rotating magnetic field produced by what became known as a "god helmet" and found that they could artificially create the experience of religious feelings—the helmet's wearer reports being in the presence of a spirit or having a profound feeling of cosmic bliss (Hitt 1999; Ruttan et al. 1990). Persinger claimed that about eight in every ten volunteers reported quasi-religious feelings when wearing his helmet (Connor 2009). Persinger's results have been widely criticized, but a technique called "transcranial magnetic stimulation" using much stronger pulsed magnetic fields has shown diagnostic and therapeutic potential in the central nervous system with a wide variety of disease states in neurology and mental health. This research is still evolving (Rossini and Rossi 2007; Ziemann 2017).

Later studies found that several areas of the brain are indeed involved in religious belief, one within the frontal lobes of the cortex—which are unique to humans—and another (the thalamus) in the more evolutionarily ancient regions deeper inside the brain, which humans share with apes and other primates. Professor Jordan Grafman, from the U.S. National Institute of Neurological Disorders and Stroke in Bethesda, Maryland, notes that "religious belief and behavior are a hallmark of human life, with no accepted animal equivalent, and found in all cultures," and further, "our results are unique in demonstrating that specific components of religious belief are mediated by well-known brain networks, and they support contemporary psychological theories that ground religious belief within evolutionary-adaptive cognitive functions" (Connor 2009). Other studies involving the neurochemistry of spirituality point to a family of brain enzymes also associated with religious experience (Mohandas 2008). Additionally, certain genetic polymorphisms have also been correlated with spirituality (Hamer 2005; Muller 2008).

Subhan Allah—Manifesting Angels

Yassir is Muslim and so we have, of course, discussed various aspects of Islam. In one of these discussions, he mentioned to me how a certain phrase in Arabic has long been associated with the presence of angels. Yassir explained that this angel-making protocol has been around for more than a thousand years and that, like Tibetan prayer flags, it makes no difference whether you speak the phrase, write it, or cause it to be printed. Any time the phrase is iterated in any form, you get an angel.

The phrase is "*Subhan Allah*" (سبحان الله), which translates into English language as the word "hallelujah" meaning "glory be!" or "praise the lord!"—a word that has come to have both religious and more secular meanings. "*Subhan Allah*" is part of daily prayers in Islamic tradition, so it is possible to roughly estimate the number of times it has ever been repeated: 100 "*Subhan Allah*'s" per day since the death of the prophet Muhammad in 632 CE × 10 billion Muslims × 60 person-years = 20 quadrillion. That's many more angels than human souls since only ~100.8 billion people have ever died.[3]

We know that numbers of molecules involved in the processes of molecular biology are significantly larger. About 10,000 gigabytes, or 10 terabytes, can be stored in a smear of DNA at the end of a small test tube, the size of the point of a pencil. One gram of DNA can hold at least one billion terabytes (one zettabyte), or 8,000,000,000,000,000,000,000 bits of data. It occurred to me, therefore, that with the application of a few of the tools of molecular biology, it might be possible to change the entire demographic of heaven. The director of my Harvard laboratory, geneticist and molecular engineer George M. Church, calculated that if we were to spend $900 on reagents for PCR, we could create 50 quadrillion angels.

This seemed to me to be a fascinating project to undertake in quarantine. Throughout spring and summer 2020, and into the early weeks of 2021, access to laboratories at both MIT and Harvard was completely curtailed or extremely limited, and all my laboratory projects were abruptly halted. Technically, I have been allowed halftime access to the lab since, but other constraints, including my particular COVID risk group, and unrelated medical conditions, also limited my access to the laboratory. To some extent, many of our laboratory vendors continued to operate, however, and though also limited by effects of the pandemic, services such as DNA synthesis and sequencing continued to be available with transport organized by post or courier. Remote access to libraries and archives had been available throughout, and many of my colleagues suddenly found themselves with more time on their hands for discussion and creative pursuits than would have ordinarily been available to them in the past. Thus, the project soon involved several others, including Sarah Khan, an artist and biologist from Peshawar, Pakistan.

One of the first problems was finding references to the "*Subhan Allah*" angel-making tradition. We found published accounts in manuscripts dating back several centuries. One, by Ibn Nur al-Din, al-Abbas,[4] recounted sources of the tradition dating to the ninth century. The practice is also referred to in Hadith collections, accounts from verbal and physical teachings and traditions dating from the early Islamic era. Although these accounts have been contested and have not been uniformly endorsed by many religious scholars, persistent narratives about pronouncements of *Subhan Allah* and the presence of angels have endured for many hundreds of years.

DNA Manifolds

The first part of our strategy to convert the words *Subhan Allah* into DNA was to convert the Arabic language phrase into Arabic language ASCII format. ASCII, abbreviated from *American Standard Code for Information Interchange*, is a character encoding standard for electronic communication. ASCII codes represent numbers and textual characters as binary information in computers, telecommunications equipment, and other devices. ASCII character-to-binary conversions require eight binary "bits" per encoded character, so that the nineteen Arabic characters in *Subhan Allah* can be expressed as a 152-bit binary number. Converting these binary "bits" into DNA calls for another encoding step.

The variable parts of DNA molecules are the sequence of structures that make up the "rungs" of their ladder-like conformations. These rung-like structures are called "bases," and DNA bases are most often referred to by their initials.

Cytosine is C
Thymine is T
Adenine is A
Guanine is G

The 4 DNA bases have four different molecular weights, and so we can increment them according to mass such that:

C = 00
T = 01
A = 10
G = 11

I have used this same strategy for encoding binary information into DNA since my early DNA-encoding projects in the 1980s (see, for example, Davis 1996). The 152-bit *Subhan Allah* ASCII code can thus be converted into the 76-mer DNA sequence shown here:[5].

GTACAGCGGTACAAACGTACAAGTGTACAATGGTATTAACCACCG TACAATGGTATACTCGTATACTCGTATACTG

This is a very efficient coding scheme that contains two binary bits in each DNA base. However, there are problems. Repeats of DNA bases in a sequence and embedded symmetries can interfere with DNA synthesis and sequencing. To date, information density of two bits/DNA base has been considered theoretically possible, but when taking into account inevitable DNA reading and writing errors, a maximum of 1.8 bits of data per nucleotide of DNA has been cited as the practical limit. For perspective, information density achieved with "*DNA Fountain*" encoding, one of the most efficient DNA data-encoding methods to date, was 1.57 bits/base (Erlich and Zielinski 2017).[6]

In any case, we didn't feel that such straightforward encoding was appropriate here. Instead, we thought we might use this opportunity to advance the practice of DNA information keeping. We were inspired by the intricate geometries and calligraphies of Islamic art to come up with a similarly elegant encoding scheme that could hold our *Subhan Allah* data in several different, but simultaneous layers of informational symmetry. For this, we decided to use a technique we call *DNA Manifolds*.

Nature uses sixty-four different triplets of DNA bases to code for the twenty amino acids that comprise all of the proteins that make up living things. There are only twenty "universal" amino acids, but there are sixty-four ways to make triplets from the four DNA bases. These sixty-four triplets are called codons. Owing to this disparity in numbers, nature generally codes for each of the twenty amino acids with multiple codons so that in almost every case, each amino acid can be represented with anywhere from two to six different codons. The relationship of the twenty amino acids to corresponding codons is what biologists call the genetic code. When there's a mutation that alters DNA bases but not corresponding amino acids, this is called a silent mutation. The first part of the *DNA Manifolds* strategy involves *Silent Code*, a method that we devised[7] for DNA-encoding using silent mutations to hold binary information in redundant codons.

The graphic icons in Figure 5.1 represent amino acids. These graphics were for me preferable to interpreting the standard abbreviations for amino acids in text form. Values are assigned to individual codons according to the respective incremental mass of all codons translating for a particular amino acid. Since even a relatively short protein can be translated from an astronomically large number of distinctly different sequences of DNA, *Silent*

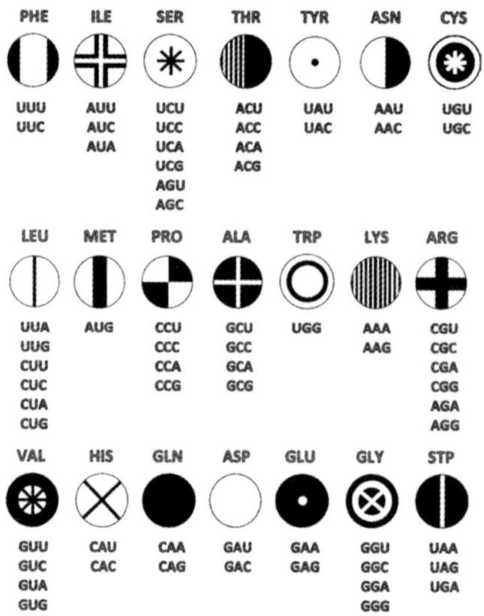

FIGURE 5.1A *Joe Davis.* Graphical Representation of 64 codons and corresponding amino acids (genetic code), *2021. Courtesy of Joe Davis.*

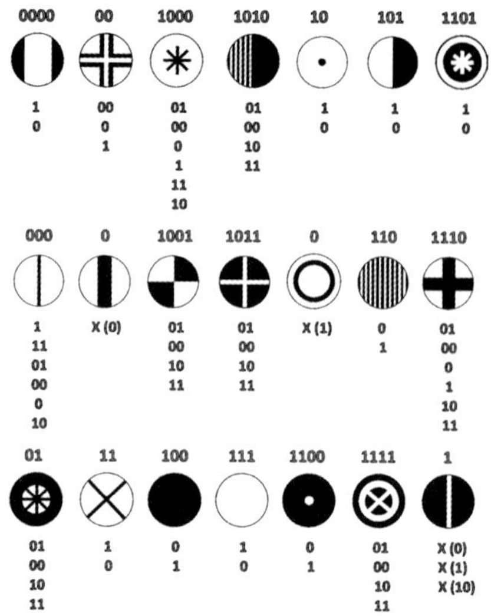

FIGURE 5.1B *Joe Davis.* Graphical Representation of Amino Acids which have been assigned distinct values *(amino code), 2021. Courtesy of Joe Davis.*

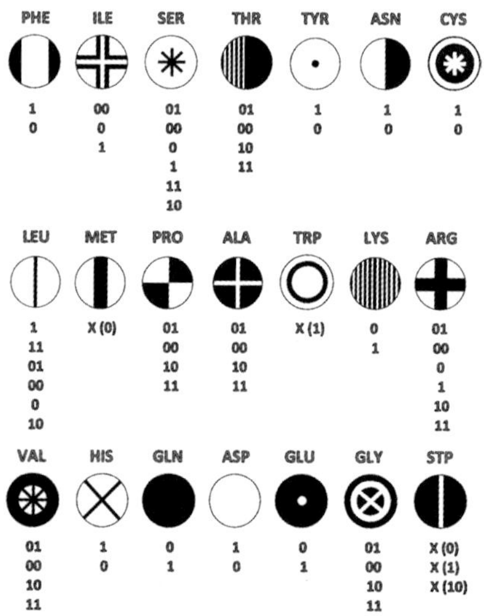

FIGURE 5.1C *Joe Davis.* Graphical Representation of Amino Acids with Values Assigned According to Mass *(silent code), 2021. Courtesy of Joe Davis.*

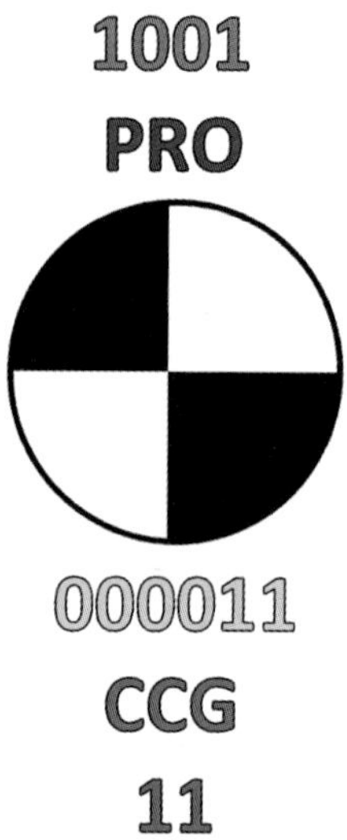

FIGURE 5.1D *Joe Davis.* Three numbers embedded within DNA Molecules: Amino (1001), Codon (11) and DNA (000011), *2021. Courtesy of Joe Davis.*

Code can be used to encode any number within a given gene or set of genes that comprise given sets of amino acids.

If amino acids are also given distinct values, then *Subhan Allah* can be coded for in a molecule that simultaneously codes for something else. A message can be independently written into a number assigned to a sequence of amino acids (*Amino Code*) irrespective of information written into the number associated with a corresponding sequence of codons (*Silent Code*). This is a very flexible coding technique, since even in the case of relatively small genes, astronomical numbers of distinctly different DNA sequences can code for the same sequence of amino acids. Nature has built functional redundancy into the genetic code, but there is some nonfunctional redundancy that is up for grabs.[8]

Within *DNA Manifolds*, there is a number for codons and a number for amino acids—but there is a third number, too—one that corresponds with the DNA sequence itself, where C =00, T=01, A=10, and G=11 (see Figure 5.1d). Thus, every DNA molecule larger than a 2-mer can hold three arbitrary numbers or what can be described as three "pages" of information, and it seems that nature uses only two of them. One page is the DNA itself; one page is the protein it codes for, and the third page is up for grabs. This "third page"—the number associated with codons—is what we call *Silent Code*, because we can use this extra page to encode information without increasing the length of the host molecule, or altering the nature of the protein.

Line one: 1101 1000 1011 00 1111 01 1000 1010 1000 . . .
Line two: CYS SER ALA ILE GLY VAL SER THR SER . . .
Line three: TGT TCG GCT ATA GGC GTT TCT ACA TCT . . .
Line four: 1 1 01 1 00 01 01 10 01 . . .

Line five: (the "third number"): 01110101001111000110 . . .

[**WHERE** T= 01, G= 11, T= 01, T= 01, C=00, G=11, G=11, C=00 . . .]

We have used two of these three "numbers" to encode the same *Subhan Allah* phrase. Both numbers are automatically contained in the DNA sequence shown in line three, which when converted to binary becomes the "third number" of the molecule (line five).

This number represents the DNA bases and it holds all of the information coded into the other two numbers, including the specific sequence of the initially encoded 76-mer DNA molecule.

These three pages of information (inherent to almost all DNA molecules) are key to the coding method we have called *DNA Manifolds*. We have shown how information can be written over itself again and again in the same DNA molecule. But importantly, we can use one of the available pages of information to code not just input data, but the map of an entirely different molecule, a virtual molecule that is entirely hypothetical, but that is nevertheless precisely described. His "third" number can then be encoded into the *Amino Code* and *Silent Code* numbers of yet another DNA molecule, and so on, cascading the original information (in this case *Subhan Allah*) into many layers of encoded information.

So, to review, a DNA molecule has what can be described as three pages of information. One page is the DNA molecule itself (expressed as codons), one page is the protein it codes for (expressed as amino acids), and the third page is the map of another DNA molecule, a "virtual" molecule that is precisely described. This virtual molecule has three pages, too: one is the molecule itself, one of its other two pages holds arbitrary input data, and its third page holds the map of a second "virtual" molecule. The second virtual molecule holds yet another three pages, too, and I could go on, but I only have a few pages in this chapter to work with here!

Since *DNA manifolds* are all about numbers, I can point out that manifolds don't have to be made of DNA. The same mathematical operations can be used to construct manifolds visually. I have, for instance, similarly constructed manifolds based on sets of four different colors; a specially-designed deck of sixty-four playing cards (with four suits), and in another example, with hundreds of 20 ml falcon tubes filled with four different quantities (levels) of sand. With, say, a piano having sixty keys, musical manifolds are possible, too, where musical compositions can be written inside of other musical compositions.

In the case of our angel-making manifold, the *Subhan Allah DNA Manifold* in Figure 5.2 is shown as eighty-six respective DNA amino acids annotated in red with *Subhan Allah* binary data in the form of *Amino Code*. The corresponding DNA codons that hold *Silent Code* are also shown in blue, and these code for the identical *Subhan Allah* binary data. Note that the numbers in blue and red are the same number.

George Church suggested yet another numerical modality for encoding Arabic language. Numerological systems of abbreviation are common among Semitic languages including Arabic and Hebrew. In Arabic, this numerological vocabulary is called *abjad*, and by this method, a numerical abbreviation is given from which the initial word can be interpreted.

The abjad number of سبحان اللهه is 187

= (5+30+30+1) + (50+1+8+2+60)

= 11 10 11 10 [binary]

= GAGA

[Lady Gaga should love this outcome!]

As the *Amino Code* requires less space than the *Silent Code* to encode the "*Subhan Allah*" binary phrase, the final thirty-six codons of the *Amino Code*

SUBHAN ALLAH MANIFOLD ENCODING:

Amino Code:	1101	1000	1011	00	1111	01	1000	1010	1000	110	1100	01	01	01	1011
Amino acid:	CYS	SER	ALA	ILE	GLY	VAL	SER	THR	SER	LYS	GLU	VAL	VAL	VAL	ALA
DNA:	TGT	TCG	GCT	ATA	GGC	GTT	TCT	ACA	TCT	AAG	GAG	GTA	GTG	GTC	GCT
Silent Code:	1	1	01	1	00	01	01	10	01	1	1	10	11	00	01

Amino Code:	1011	000	1010	01	111	1011	00	101	1010	000	01	000	00	1101	1000	1010
Amino acid:	ALA	LEU	SER	VAL	ASP	CYS	ILE	ASN	THR	LEU	VAL	LEU	ILE	CYS	SER	THR
DNA:	GCT	CTT	TCC	GTT	GAT	TGC	ATA	AAT	ACC	CTT	GTT	CTT	ATA	TGC	AGT	ACA
Silent Code:	01	01	00	01	1	0	1	1	00	01	01	01	1	0	11	10

Amino Code:	01	11	11	01	1001	1000	01	00	11	01	1001	1000	01	00	1101	1001
Amino acid:	VAL	HIS	HIS	VAL	PRO	SER	VAL	ILE	HIS	VAL	PRO	SER	VAL	ILE	CYS	PRO
DNA:	GTG	CAC	CAC	GTT	CCT	TCC	GTG	ATA	CAT	GTA	CCG	TCC	GTA	ATA	TGT	CCT
Silent Code:	11	0	0	01	01	00	11	1	1	10	11	00	10	1	1	01

Amino Code:	1000	01	11	–	111	01	110	1110	1110	1110	1110	1110	1110	1110	1110
Amino acid:	SER	VAL	HIS	–	ASP	VAL	LYS	ARG	ARG	ARG	ARG	ARG	ARG	ARG	ARG
DNA:	TCC	GTC	CAC	–	GAT	GTC	AAA	CGC	AGG	CGT	AGA	CGC	AGA	AGA	CGT
Silent Code:	00	00	0	–	1	00	0	00	11	01	10	00	10	10	01

Amino Code:	1110	1110	1110	1110	1110	1110	1110	1110	1110	1110	1110	1110	1110	1110
Amino acid:	ARG	ARG	ARG	ARG	ARG	ARG	ARG	ARG	ARG	ARG	ARG	ARG	ARG	ARG
DNA:	AGG	AGG	CGT	AGA	CGT	AGA	CGC	CGT	CGC	AGG	CGT	AGA	CGT	AGA
Silent Code:	11	11	01	10	01	10	00	01	00	11	01	10	01	10

Amino Code:	1110	1110	1110	1110	1110	1110	1110	1110	1110	1110	1110
Amino acid:	ARG	ARG	ARG	ARG	ARG	ARG	ARG	ARG	ARG	ARG	ARG
DNA:	CGC	CGT	CGC	AGG	CGT	AGA	CGT	AGA	CGC	CGT	AGG
Silent Code:	00	01	00	11	01	10	01	10	00	01	11

FIGURE 5.2 Subhan Allah DNA Manifold. *Joe Davis. 2021.*

in the *Subhan Allah DNA Manifold* is used to encode 17.5 "*Subhan Allah*" abjad ("GAGA") repeats.

The *Subhan Allah DNA Manifold* that contains all this information is a 258-mer DNA molecule, which is represented in DNA as:

> TGTTCGGCTATAGGCGTTTCTACATCTAAGGAGGTAGTGGTCGCTGCTCTTTCCGTT
> GATTGCATAAATACCCTTGTTCTTATATGCAGTACAGTGCACCACGTTCCTTCCGTGA
> TACATGTACCGTCCGTAATATGTCCTTCCGTCCACGATGTCAAACGCAGGCGTAGAC
> GCAGAAGACGTAGGAGGCGTAGACGTAGACGCCGTCGCAGGCGTAGACGTAGACG
> CCGTCGCAGGCGTAGACGTAGACGCCGTAGG

Our example is reasonably efficient in terms of maximizing the number of bits that can be stored per DNA base. In this case, 2 × 152-bit *Subhan Allah* texts, in binary Arabic ASCII, are encoded, as well as 17.5 × 8-bit abjad "GAGA" encodings, totaling 444 bits in 258 DNA bases, or 1.72 bits/

DNA base. If the two encoded 76-mer DNA sequences and the 70-mer DNA encoding the 17.5 "GAGA" repeats are also counted, then input data totals 748 bits in 258 DNA bases, or 2.89 bits/DNA base. As previously noted, information density with DNA Fountain encoding, one of the most efficient DNA data-encoding methods to date, achieved 1.57 bits/base. So far as we know, our *Subhan Allah DNA Manifold* demonstrates greater information density than any other DNA information-keeping strategy published to date. This is how we can hold 2.417 hundred million billion angels (or Subhan Allah iterations) on the .75 mm pinhead of an average straight pin.

We added DNA linkers to our sequence at the terminal ends so that the final design called for a 301-mer DNA. These linkers facilitate integration with a plasmid—think of a plasmid as a kind of bus used to ferry exogenous DNA passengers into bacterial cells—and also happens to be standard prefixes and suffixes used by the iGEM community to create standardized "biobricks."[9] We hope that standardizing the *DNA Manifold* in this way will help to enable worldwide distribution of the *Subhan Allah* sequence.

GeneUniversal, a biotech company in Newark, Delaware, was able to commercially synthesize and sequence our *Subhan Allah DNA Manifold* . They were also able to insert it into pUC57, a plasmid bacterial vector. The total cost of commercial synthesis of the *Subhan Allah DNA Manifold* and its assembly within the plasmid vector was less than $110. This is remarkably economical, since synthesis alone would have cost five or ten times this amount a few short years ago.

Our collaborator, Kyle Cromer, resequenced the *Subhan Allah DNA Manifold* to confirm its accuracy at Stanford University. Using a simple process known as transformation, the DNA plasmid can be added to *E. coli* bacteria, which proliferate reproducing the plasmid inside them. We have since grown and harvested copies and have been distributing them to laboratories worldwide. *Subhan Allah DNA Manifold* plasmids arrived at Church Lab in the midst of the pandemic and were intercepted by my colleague Gabriel Filsinger. These plasmids were also transformed with *E. coli*, and colonies have been grown to harvest more plasmids at Harvard Medical School. Now our angels are on the loose! Batches of *Subhan Allah* DNA have arrived at Stanford University, Cold Spring Harbor Laboratories, University of Kentucky, Carnegie Mellon University, Harvard—and Sarah Khan now has them in Pakistan.

Conclusion

Yet another chapter is about to be added to the story of *DNA Manifolds*. I have mentioned how I assigned numbers to DNA bases according to molecular weight. I also used molecular weight as a guide for the *Silent Code* number assignments, and when I assembled all the manifold number sets together with corresponding molecular weights, I noticed something peculiar. The genetic code contains another group of heretofore unnoticed redundant number sets. There are five nonredundant codons in the traditional genetic code (three "stop" codons, and methionine and tryptophan, the only amino acids represented by single codons). In what seems to be an odd coincidence, there are also five similarly nonredundant molecular weights in the code but these differ from the nonredundant codons in the traditional code. I organized groups of codons associated with these redundant sets of molecular weights and assigned them with yet another set of binary number values, thus generating a fourth page for *DNA Manifolds*. Sarah Khan and I have recently begun to explore the idea of a new project to demonstrate practical applications of four-page manifolds. It will be poetic or culturally resonant, of course, given that we are artists after all.

I suspect there may be other pages, too. *DNA Manifolds* seem to have revealed a special kinship between numbers and I wonder if there is some hidden symmetry among them that we all seem to have missed, a hidden symmetry in mathematics that has already been co-opted by natural biological processes. One possibility would be redundancy in absorption spectra of DNA codons at different wavelengths of light, or even at different frequencies of sound. It is likely that other new pages will be discovered with the potential to revolutionize efficiency and practical applications in many aspects of mathematics as well as in DNA information keeping.

What at first seemed to be a simple poetic gesture has become much more complex. We have explored new forms of information keeping in DNA and we have pioneered new expressions of Islamic art in biological form. We have been building bridges between art and mathematics, between art and science, science and spirituality, and there is also the not-so-subtle point that we may in some small part have helped to mitigate widespread fear and hatred of Islam. While we have lost over 900,000 lives to the pandemic in the United States alone (as of May 2021), the notion that we may be helping to create angels could come as a comfort and solace to many whose lives have been disrupted by the terrible impacts of COVID-19.

Notes

1 Uncovered in 1957–61 by Ralph Solecki and his team from Columbia University, this yielded the first adult Neandertal skeletons in Iraq, dating between 60 and 80,000 years bp.

2 Species traditionally used as diuretics, stimulants, astringents, and anti-inflammatories—such as yarrow, cornflower, ragwort, groundsel, horsetail, and hollyhock—were represented in pollen samples.

3 100,825,272,791 according to Carl Haub, a senior demographer at the Population Reference Bureau (Population Reference Bureau n.d.).

4 We were able to obtain a nineteenth-century copy of the Ibn Nur al-Din, al-Abbas reference (republished narratives from 1735) from a bookseller in Cairo (*Nuzhat al-jalis wa munyat al-adib al-ani* 1876).

5 -mer from the Greek *meros* or part is used to denote the number of nucleotides in an RNA or DNA sequence. Nucleotides are comprised of one of the DNA/RNA nucleobases: Adenine (A), Guanine (G), Thymine (T), Cytosine (C) in DNA, and Uracil (U) in RNA.

6 Fountain codes are a class of practical erasure-correcting codes first invented by Michael Luby in 1998 and published in 2002. These codes were developed to cope with erasure prone channels, such as the internet, where files are transmitted in multiple small packets each of which is either received without error or not received at all. By 2017, fountain codes were already in use to cope with imperfect digital memory storage and recovery and information transfer over noisy broadcast networks, such as wide-bandwidth phone lines and satellite communications systems.

7 Ideas for *Silent Code* evolved from my discussions with Dana Boyd in the 1990s; see Davis et al. (2006).

8 In addition to having information of its own, values assigned to amino acids (peptide sequences) may also be used as a check for copying errors. A predictable sequence of amino acids can hold core information while its triplet variants can encode separate data sets. Methods for such over-encoding of information are common aspects of electronic and broadcast communications where multiple layers of information are added to guarantee the integrity of information sent or received.

9 The term "iGEM" is short for international Genetically Engineered Machine. The phrase arose originally from the competition of the same name, which encourages high school and university students around the world, from all disciplines, to come up with innovative and standardized methods of genetically modifying living organisms for the advancement of humanity.

References

Connor, S. (2009). "Belief and the Brain's God Spot." *The Independent*. https://www.independent.co.uk/news/science/belief-and-the-brain-s-god-spot1641022.html

Davis, J. (1996). "Microvenus." *Art Journal* 55(1): 70–4. doi: 10.2307/777811

Davis, J., Boyd., D., O'Reilly, H., and Wieczorek, M. (2006). "Art and Genetics." In *eLS* (ed.). doi: 10.1002/9780470015902.a0005868

Erlich, Y., and Zielinski, D. (2017). "DNA Fountain Enables A Robust and Efficient Storage Architecture." *Science* 255(6328): 950–4.

Hamer, D. H. (2004). *The God Gene: How Faith is Hardwired into Our Genes*. Doubleday.

Hitt, C. (1999). "Toward an Ecological Sublime." *New Literary History* 30(3): 603–23. https://www.jstor.org/stable/20057557

Ibn Nur al-Din, al-Abbas (1876). *Nuzhat al-Jalis wa Munyat al-Adib al-Anis. Al-Matba'a al.* Wahbiyya Publishers. Vol. 1, p. 287.

McNamara, P. (2009). *The Neuroscience of Religious Experience*. Cambridge University Press. doi: 10.1017/CBO9780511605529

Mohandas, E. (2008). "Neurobiology of Spirituality." *Mens Sana Monographs* 6(1): 63–80. doi: 10.4103/0973-1229.33001

Muller, R. J. (2008). "Neurotheology: Are We Hardwired for God?" *Psychiatric Times* 25(6): 24. https://link.gale.com/apps/doc/A180317268/AONE?u=anon~346a7833&sid=googlecholar&xid=0d7a7fe8

Rossini, P. M., and Rossi, S. (2007). "Transcranial Magnetic Stimulation: Diagnostic, Therapeutic, and Research Potential." *Neurology* 68(7): 484–8. doi: 10.1212/01.wnl.0000250268.13789.b2

Ruttan, L. A., Persinger, M. A., and Koren, S. (1990). "Enhancement of Temporal Lobe-Related Experiences During Brief Exposures to Milligauss Intensity Extremely Low Frequency Magnetic Fields." *Journal of Bioelectricity* 9(1): 33–54. doi: 10.3109/15368379009027758

Ziemann, U. (2017). "Thirty Years of Transcranial Magnetic Stimulation: Where Do We Stand?" *Experimental Brain Research* 235(4): 973–84. doi: 10.1007/s00221-016-4865-4

6

Gene Music

Biologically Motivated Musical Serialism

Ira Fleming

Musical notation takes on new and complex meaning when played out loud. Genetic sequences of DNA and RNA, similarly, take on new form and meaning when translated into amino acids, proteins, and the complexity of life itself. This similarity in relationship between a code and its expression has prompted a few musicians and scientists to convert genetic sequences into musical patterns, creating what will be referred to in this essay as "Gene Music." The gene-to-music conversion outlined in the next few pages involves the assignment of notes on the 12-tone scale to amino acids, the building blocks of protein, rather than DNA bases that encode those amino acids. Dissonance or varied note duration are added to the composition based on the degree of conservation of genetic information across species, a proxy for evolutionary selection. This approach to Gene Music, using patterns embedded within amino acids, creates musical complexity, and the focus on conservation of sequence highlights strings of amino acids/notes that are critical to protein function. The music is bizarre and illuminating. Playing or hearing it yields a nonrandom aesthetic experience rooted in the form and function of proteins. As advances in genetics are made and the manipulation of DNA sequences becomes more common, Gene Music reminds us that biological sequences are pervasive and essential to form, function, and meaning.

History of Gene Music

Composers have long recognized that sequences can be mined for musicality. One influential approach is serialism, a technique using a set of values or an algorithm to produce musical compositions, developed in the early twentieth century by Arnold Schoenberg and his collaborators (Hyde 1985). Schoenberg's 12-tone technique involves the production of a tone row, a series in which each note in the chromatic scale is used only once. The composer may then subject the tone row to replications, inversions, and other manipulations, resulting in a new composition. Notably, these transformations may also occur in genetic sequences. John Cage went further, elevating sequences created by chance into musical compositions. His *Music of Changes* used "chance operations" sourced from the *I Ching*, a classic Chinese divination text, to produce randomized musical outputs, ceding human judgment about aesthetic value (Jensen 2009).

Susumu Ohno saw that nucleotides in DNA (commonly represented by letters A, T, C, and G) could be used as predetermined values in composing serialist music, transforming the medium of naturally arising patterns. Two notes in a scale were assigned to each nucleotide, with lower frequency notes corresponding to nucleotides with greater atomic weight. Ohno translated part of the gene for a certain human protein essential to metabolism into a composition in D minor, noting that the resulting tune sounded "hauntingly melancholy, as though reflecting the Weltschmerz of the gene that persevered for hundreds of millions of years" (Ohno 1986). Ohno composed other works that highlighted different features of biological sequences, including palindromic sequences of amino acids within proteins (Ohno 1993). Since then, others have produced similar works with various methodologies such as pairing amino acids, rather than nucleotides, with notes or chords, or even implementing artificial intelligence to develop novel gene-like music from a dataset of compositions (Yu et al. 2019). Some Gene Music composers search to break down aesthetic norms, others to highlight features of genetics such as repetition, and some claim that music made from gene sequences can make the study of genetics more accessible to children or the visually impaired (Takahashi and Miller 2007).

Musicality of Genetic Sequences

My method of composition highlights three aspects of genetic and musical sequences: (1) high-order nonexact repetition of sequences; (2) low-order exact repetition of simple sequences or single notes/amino acids; and (3) informational content. Repetition, both high- and low-order as in genetics, is pervasive in musical compositions. The repeated strike of a drum, a "steady beat," constitutes repetition. More complex patterns are built over this base, which themselves are repeated. As Eugene Narmour argues in his

Implication-Realization model, even if music lacks repetitive qualities, the *expectation of* repetition shapes the listener's experience (Narmour 1989).

The human brain is an exceptional tool for recognizing high-order repetition and patterns, especially in music. Though it is unlikely that someone listening to Gene Music would be able to identify the underlying presence of an alpha helix structure, there are simpler repetitive regions in gene sequences that, when translated to music, create a pattern the listener recognizes. These repeated sequences mutate and evolve to varying degrees, becoming variations on a theme that could be audibly picked out of a Gene Music composition. Single amino acid repeats or other low-order repetition may also be identified. For example, the number of "CAG" trinucleotide repeats in the Huntingtin gene is inversely related to survival time in Huntington's disease and would likely be recognized after sonification of the sequence (Langbehn et al., 2010). Informational content is more difficult to define than repetition given that, in one case, the information involves emotional response to melody, and in the other, it involves the function of proteins, though both kinds of information are reliant on the "translation" of the sequence into a new medium.

One might predict that gene sequences would have musical quality given the presence of repetition and informational content. In Figure 6.1, this shared characteristic of repetition is visualized, with both the solo violin melody from the first movement of Vivaldi's Violin Concerto in A minor and the amino acid sequence of the human Slit 1 protein, a driver in development. They are shown side by side in the form of a dot-plot, where "similarity" across different parts of the piece or gene are scored via identity scoring matrix, with higher scores shown as dark pixels. Both the musical composition and amino acid sequence appear highly structured given their

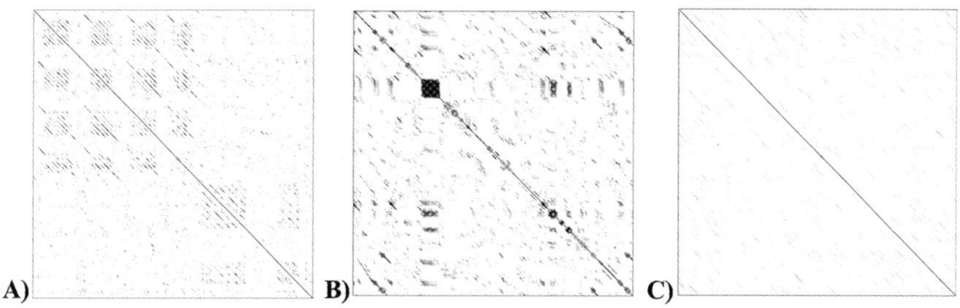

FIGURE 6.1 *Ira Fleming, 2023, Dot-plots for (A) human Slit 1 protein, (B) Vivaldi's Violin Concerto in A Minor Op. 2 No. 6 Mvt. I Allegro, and (C) a randomly generated amino acid sequence. Courtesy of Ira Fleming.*

Alignment scores were produced via identity scoring matrix and images were generated using the Dotlet software with a window size of 15 residues/notes (Junier and Pagni 2000). X and Y axes represent the identical sequences. Each pixel corresponds to a locus on the horizontal and vertical sequences and is darker if that locus is highly similar.

repetition (see Figures 6.1a and 6.1b). In contrast, Figure 6.1c is a randomized sequence of amino acids which displays no obvious structure.

Method of Composition

I. Primary Melody

In my method of Gene Music composition, the primary melody is dictated by the consensus sequence of amino acids in a protein. To prevent information loss, one amino acid must correspond to only one note every time it appears in the polypeptide sequence. The problem arises that there are twenty commonly used proteinogenic amino acids and only twelve notes in the Western Chromatic Scale. To circumvent this problem and allow for bi-directional conversion, the same notes in different octaves are considered different notes. A preexisting melody and an amino acid sequence serve as the inputs. The most common note in the melody is assigned to the most common amino acid in the protein. Then, the most common precursor to that amino acid is assigned to that of the note, and so on. This continues until the program encounters a note or amino acid that has already been assigned, whereupon it will pair the next most common amino acid and note until every amino acid is assigned a note.

II. Secondary Characteristics

One more feature of the genetic code highlighted, which has no obvious analogue in music, is the evolutionary conservation of sequences. Mutations in genes are ubiquitous and the driving force in evolution. Certain domains of gene sequences, however, are preserved without frequent mutation in populations because of an evolutionary incentive to maintain a region of the protein. These residues, maintained in homologous genes (genes of similar sequence with a shared ancestral origin), are regarded as "conserved," and have a greater impact on protein function when mutated.

My method reinterprets residue conservation across forty diverse genomes pulled from different species as polyphonic harmony with a degree of consonance corresponding to the degree of amino acid conservation. Conservation of residues in a gene is determined by the percent of residues that match the consensus sequence produced from homologs. Highly conserved regions are assigned more consonant harmonies to relay their structural significance. Since dissonance is complex, artistic liberties were taken arranging intervals from roughly dissonant to consonant. All the intervals are chosen according to the following rules:

<40% conservation—minor second (m2) or major seventh (M7)
40–60% conservation—major second (M2) and minor seventh (m7)

60–70% conservation—perfect fourth (P4)
70–90% conservation—minor and major thirds (m3 and M3) and sixths (m6 and M6)
90–100% conservation—perfect fifth (P5) and perfect octave (P8)

Vivaldi's Violin Concerto in A Minor Op. 2 No. 6

The solo violin part from Vivaldi's Violin Concerto in A Minor Op. 2 No. 6 Mvt. I Allegro was chosen as musical input for my first composition because of its fame, accessibility, and refined repetitive structure common within baroque compositions. In the first seven measures of Vivaldi's melody (seventy-one notes), there are five strings of low-order repetition, sequences of the same note repeated five times. The repetition of broader motifs or *ritornelli* is an integral part of the concerto's structure, incorporating the element of nonexact repetition of whole sequences. The repeated themes divided by variable sequences lends continuity to the music, leverages the relationship between familiarity and enjoyment, and subverts or reinforces the expectations of the listener depending on the passage.

The Slit Protein: Exploration of the Leucine-Rich Repeat (LRR) as a Theme

The consensus sequence of an alignment of Slit protein homologs from different species serves as the other input for the Gene Music composition. Secretory extracellular matrix proteins make up the Slit protein family, which is involved with the growth, structure, and adhesion of cellular systems, and signaling in neural development. One most identifiable repeated sequence is defined as Leucine Rich Repeats (LRRs) (Rothberg 1990).

The strongest argument to produce music using the Slit protein family was the internal repetition from these repeated domains as well as low-order repetition. The Slit protein in the *Drosophila melanogaster* fruit fly has numerous sequences of low-order amino acid repetition and the LRRs constitute the kind of regular structure that allows for effective functioning of the Slit protein and are reminiscent of the *ritornelli* in Vivaldi's composition.

Production of the Slit Protein Composition

Three regions of the Slit protein were selected to compose this piece which include generally conserved regions and variable regions, low-order repetition (e.g. NNNN), and high-order repetition (LRRs). One 18-residue region with low conservation was translated into music as well as two highly conserved corresponding 18-residue sections from LRRs as variations on the LRR theme.

Residues 38-56:	AEPYGGSSGSVGIHIPGGS
LRR 1:	LEILTLNNNNLTSLPHNAF
LRR 2:	LSLLSLYDNNIQSLANGTF

The primary melody from the Slit consensus sequence is shown in Figure 6.2a. When played aloud, the LRRs are audibly more similar to each other than they are to the other domain. The LRRs notably start on the note E and end on G which biases the listener toward hearing similarity between those two sequences. Once the harmonies corresponding to residue conservation were applied (see Figure 6.2b), the LRR domains become less recognizable as the assigned note is not weighted differently than the assigned harmony, though the first sequence is subjectively more dissonant than the LRRs.

I then paired note duration to conservation, with longer notes corresponding to higher degrees of conservation. In Figure 6.2c, the same thresholds of conservation were applied to the harmonies to produce a tune with variable note lengths ranging from 32nd notes (<40% residue conservation) to

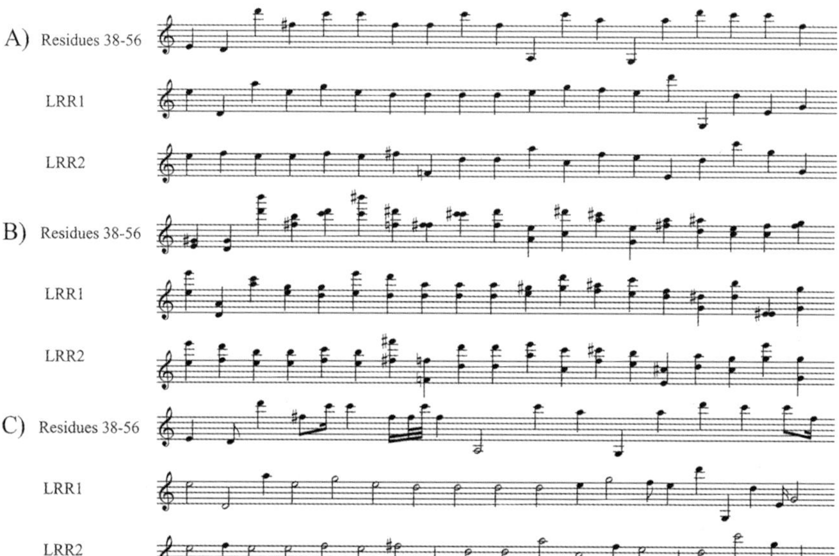

FIGURE 6.2 *(A) Melody derived from Slit protein domains, (B) Addition of polyphony with consonance of intervals corresponding evolutionary conservation for each residue, and (C) Addition of variable note durations corresponding to evolutionary conservation of residues.*

half-notes (>90% residue conservation). In these compositions, conserved domains take on an air of purpose and solemnity while variable regions are bizarre and frantic. The manipulation of note duration lends a memorable "earworm" quality to the brief melodies.

Conclusion

This attempt at Gene Music yields an aesthetic experience, and the knowledge that each note corresponds to a biological unit at the heart of life on earth gives a certain weight to the piece. Though the pattern-based sequences of music and the genome generally arise independent of each other, it is worth considering when listening to music, that patterns and repetition as well as inversions and transpositions of sequences are essential to our beings and to the course of evolution. Consider also what a genetic phenomenon like clustered regularly interspaced short palindromic repeats (CRISPR) in bacterial DNA might sound like, given the musicality of palindromes and repetition. Conversion of sequences into a new medium, like notes on a page into sound or coding DNA into RNA and protein, is required to reveal additional meaning, and perhaps some of our faculties for understanding one kind of sequence, genetic or musical, can be applied to the other. This process of Gene Music production serves as a gentle reminder that sequence, biological or musical, is fundamentally universal.

References

Hyde, M. M., and Schoenberg (1985). "Musical Form and the Development of Schoenberg's 'Twelve-Tone Method'." *Journal of Music Theory* 29(1): 85–143.

Jensen, M. G. (2009). "John Cage, Chance Operations, and the Chaos Game: Cage and the 'I Ching.'" *The Musical Times* 150(1907): 97–102.

Junier, T., and Pagni, M. (2000). "Dotlet: Diagonal Plots in a Web Browser." *Bioinformatics* 16(2): 178–9. https://github.com/sib-swiss/react-dotlet

Langbehn, D. R., et al. (2010). "CAG-Repeat Length and the Age of Onset in Huntington Disease (HD): A Review and Validation Study of Statistical Approaches." *American Journal of Medical Genetics. Part B, Neuropsychiatric Genetics: The Official Publication of the International Society of Psychiatric Genetics* 153B(2): 397–408.

Narmour, E. (1989). "The 'Genetic Code' of Melody: Cognitive Structures Generated by the Implication-Realization Model." *Contemporary Music Review* 4(1): 45–63.

Ohno, S. (1993). "A Song in Praise of Peptide Palindromes." *Leukemia* 7(2): S157–9.

Ohno, S., and Ohno, M. (1986). "The All Pervasive Principle of Repetitious Recurrence Governs Not Only Coding Sequence Construction but Also Human Endeavor in Musical Composition." *Immunogenetics* 24(2): 71–8.

Rothberg, J. M., Jacobs, J. R., Goodman, C. S., and Artavanis-Tsakonas, S. (1990). "Slit: An Extracellular Protein Necessary for Development of Midline Glia and Commissural Axon Pathways Contains Both EGF and LRR Domains." *Genes & Development* 4(1): 2169–87.

Takahashi, R., and Miller, J. H. (2007). "Conversion of Amino-Acid Sequence in Proteins to Classical Music: Search for Auditory Patterns." *Genome Biology* 8(5): 405.

Yu, C. H., et al. (2019). "A self-consistent sonification method to translate amino acid sequences into musical compositions and application in protein design using artificial intelligence." *ACS nano* 13(7): 7471–82.

7

Transformation

An Exercise in How to Relate to Lively Material

Louise Mackenzie

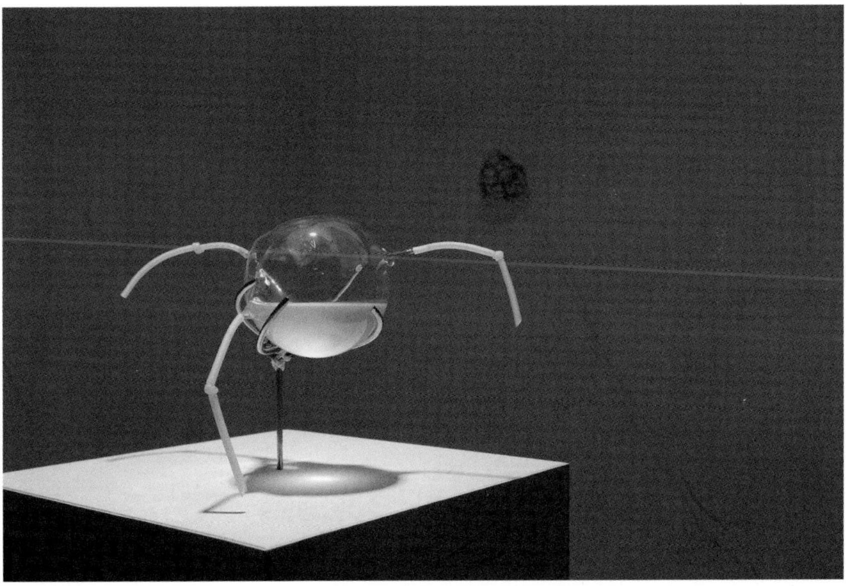

FIGURE 7.1 *Louise Mackenzie,* Pithos, *2016–. Installation view. Gallery North Project Space, Newcastle upon Tyne, UK. Image: Jason Revell. © Louise Mackenzie.*

Introduction

This chapter discusses a performative art practice that considers human/nonhuman ethical relations regarding genetic modification through an anthropomorphic filter. It offers an alternative reading of the use of DNA and *E. coli* bacteria in the laboratory to clone genetic information for the purposes of biotechnological research. Taking the reader on a guided walk through this process from the perspective of myself as artist-researcher, the chapter then recounts a form of eulogy to the lives of genetically altered laboratory organisms, charting their ancestral lineage, before concluding by reflecting upon the place of anthropomorphism in shaping our duty of care to others.

Relatable Anthropomorphism

How might we transform ourselves and in doing so transform our understanding of the other?

Anthropomorphism, while commonplace in art, is often critiqued in other disciplines as inadequate in developing our understanding of nonhuman life, and yet research within human-animal studies suggests that avoiding anthropomorphism only serves to "create other morphisms" (Karlsson 2012) and that situated anthropomorphism can bring new insights in making sense of nonhuman interactions (Servais 2018). While being within a human body may prevent me from truly understanding what it is like to be another species (Nagel 1974), it is nevertheless clear to me that I can only experience another species from my own perspective. Conversely, my understanding of the objects I encounter in the world around me is only one possible understanding of these objects (von Uexküll 1934). For example, the Luria Broth in which I grow *E. coli* bacteria may be the extent of those organisms' experience of the world, just as the liquid in our guts may form the limit of the worldview of their nonlaboratory relatives. All of these versions of experience come together in making our shared world (Schroer 2021: 135; Haraway 1992: 298) and through more-than-human activity, all share modes of passage from one to the other. It is therefore possible to explore, through a material understanding of our environment, the links between worlds and worldviews beyond those we commonly appreciate as humans.

I suggest that through anthropomorphic forms of relation, we can begin to pay attention to the affect of our technological enquiry into what I have defined as lively material (the DNA, viruses, and single cells that exist on the cusp of what we have come to term life). In *Transformation* (2017), through performing anthropomorphism, participants are guided in the act of genetically modifying laboratory *E. coli* before reflecting (to camera) upon what they have done. The following text invites you into the context of this workshop.

Psychotransgenics, or The Detournement of Cell Cloning

Take a walk with me as I stroll through the process of genetic manipulation. I will start with a brief introduction to the area. The space of cell cloning. Cell cloning is a method used in laboratories to grow and maintain segments of DNA for use in genetic research. It involves inserting custom-designed (synthetic) DNA into living bacteria (often *E. coli*) and then growing them to allow sufficient quantities of the DNA of interest to be harvested. To insert synthetic DNA within a living organism there are two key steps: the synthesis of the DNA and the process of inserting this within the host organism (known as transformation). To harvest the DNA, the bodies must perish. I learned this the hard way, and so I'm going to try to guide you through the process a little more gently.

You are going to introduce something into a body. A living body. You are going to give it some thing that wasn't there before.

First, think about what this thing could be. It could be a thought that you want to share. Perhaps it is a physical thing. Imagine that it is both things. Maybe you want to produce seasonal fruit all year round. Possibly you have thought of a way to make microbes power cars instead of petrol. You might have an idea that could end cancer or rid the world of COVID.

You are going to make your thought concrete. Exhale it as whispers in the air. Write it as words. Turn it into action. Bring it to life.

How will you do this?

Body Language

Imagine that your mind and your body understand different languages. Think of what you are about to do as starting a conversation with some body whose language you are only beginning to learn.

Now this thing that you are going to introduce, you are going to write it in this new body language. Currently, it's known as the genetic code and I'd say that in terms of language skills, we are still at the phonics stage. We love playing with the sounds (GA GA), but we don't always understand the full context in which they will be received.

So, as you are a beginner, you are going to start small. You are going to give your idea to a microbial body in a language that it can understand. All living bodies learn to read words in this language as they grow and the words become fleshy with meaning: a flagella here, a sickle cell there.

As with any language, it is shaped by context, and all bodies tell their own unique story.

Let me share with you my story of trying to read body language.

Transformation

Between 2015 and 2018, I worked with many forms of body in the laboratory of Professor Volker Straub at what was then called the Institute of Genetic Medicine at Newcastle University in the UK.

I translated a thought, a question I had, into DNA. I developed a cypher that matched codons (basic functional units of DNA) to phonemes (basic sound utterances in human speech) and worked with speech synthesis expert Étienne de Crécy to develop a program called the *Genophone* whereby any word in the English language can be translated into genetic code, and conversely, any sequence of DNA can be translated into a series of utterances, not necessarily English, but audible as speech (Mackenzie and de Crécy 2016).

It was important to me that this thought, this question I had, become more than synthesis. I made this DNA on a computer screen, then, supported by Straub's team, sent my synthetic recipe to a laboratory, which sent me back plasmid DNA. Plasmid DNA is a circular piece of DNA that floats freely within the body of bacterial cells and other cellular organisms. Let's think of it like loose change—it definitely has value, but it's not part of the household budget. It is not essential to the main functioning of the organism, but it is retained, sometimes used, and even passed onto offspring. With this plasmid DNA, I was able to change the bodies of *E. coli* bacteria. The process seems traumatic; it involves bringing *E. coli* out of -80°C cryogenic sleep and shocking them awake by placing them in warm water, which makes their bodies stretch so wide that the plasmid DNA slips through tears in the membrane of the organism. As a result, these *E. coli* now hold within them the plasmid DNA that I had designed. Dada! Through the process known as transformation, these readymade organisms now culturally and genetically held my thought within their bodies.

What did I store within them?

This question:

"What will happen if I store this thought safe within you?"

Why did I do this? Why did I burden these beings with this extraneous DNA?

I wanted to understand what it is that science does in genetically modifying other species. I wanted to go through the experience for myself and express it in my own words. Or rather, I wanted to give my idea to a body other than mine and know what it means to have my own words expressed within another body. What would transforming the organisms transform in me?

I realized that I was now responsible for these bodies in a new way. Suddenly I had progeny, and as all new parents do, I chose to name these new beings: *bioassemblages*—a biological/cultural hybrid of idea and molecular matter (Mackenzie 2016). As fast as I had brought these

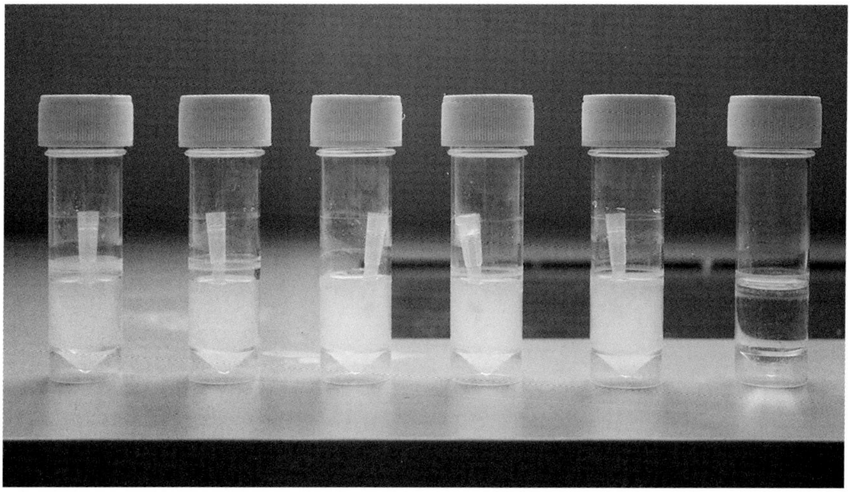

FIGURE 7.2 *Louise Mackenzie,* Lively Material, *2018. Video Still. © Louise Mackenzie.*

transformed bodies, made not born, into the world, they began to grow and multiply and I had to make decisions about how many of them I could keep looking after and how many I had to let go. I didn't find this easy, but in the end, I found my methods.

Documenting my experiences in the video diary, *Lively Material* (2018), I grew the transformed *E. coli* in plastic tubes containing a liquid nutrient called Luria Broth. The liquid turns cloudy as the number of *E. coli* increase and the available nutrients decrease. Each week, I had to select a tiny proportion of *E. coli* from the cloudy medium and place them into a fresh tube. I thanked the remainder for their service to me, said my goodbyes, and hastily took a photograph (not exactly lab protocol). Then I placed them in the bin marked "hazardous waste."

I commemorated the lives of each and every one of my bioassemblages with what I referred to as a *memento perimortem* (Mackenzie 2017). Memento is from the Latin phrase, *memento mori*, which translates as "remember (that you have to) die."[1] Perimortem is a term commonly used in clinical settings and forensic medicine. It means "occurring at the time of death or very near to it."[2] We are perhaps more familiar with the term "postmortem." Even though I felt an affinity toward the bodies that I had transformed, I was obliged to desist from keeping all of my progeny, lest they take over the laboratory. *Memento perimortem* was thus a nod to both art-historical *memento mori* and Victorian postmortem photography—a collection of images that commemorate the near death of the organisms I had bestowed my thought upon. Paying my respect in this way, allowed me to remember that the resource which I drew upon for my research was once a life.

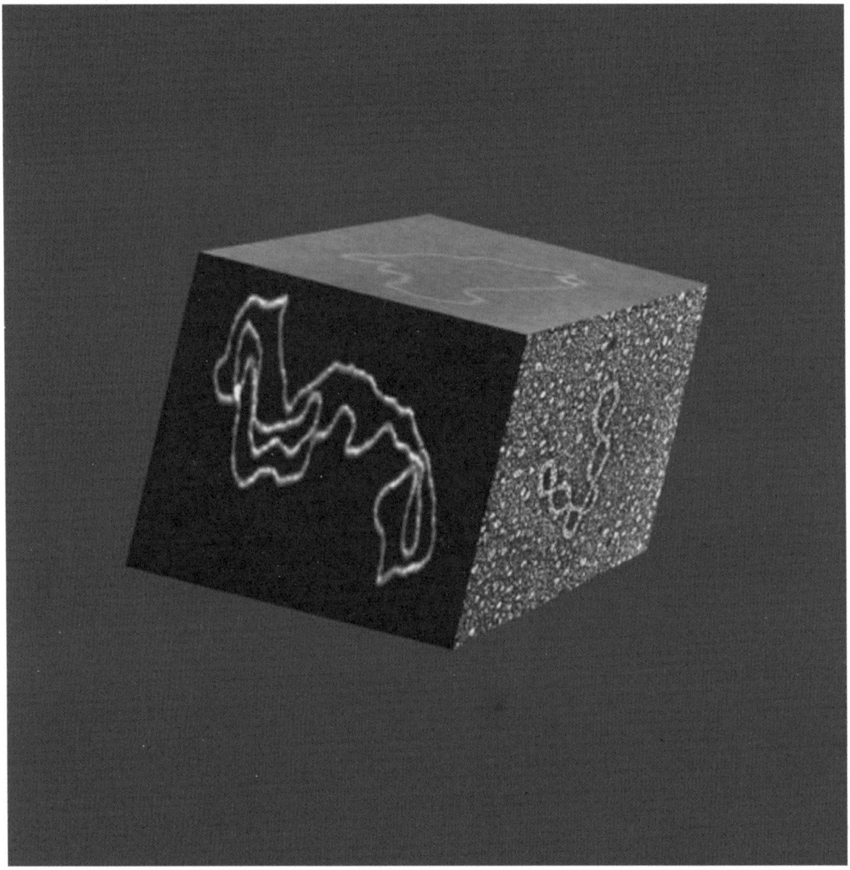

FIGURE 7.3 *Louise Mackenzie,* Flat Lively Objects, *2017. C-Type Print.* © *Louise Mackenzie.*

Ancestry of a Bioassemblage

In scientific texts, the plasmid is depicted as a circle. Yet the plasmid's form as seen through the microscope is folded and twisted. The circle reduces the plasmid to a singular function—that DNA transmits information—but see how it moves (see Figure 7.3). No matter how many angles we look from, we can only glimpse the liveliness of the material, constantly in motion and constantly relating. Plasmid DNA exists in living organisms and is also manufactured for use in research. Synthetic DNA, like all resources, has a real-world origin. Originally manufactured from fish silt, the process was made cost-effective with sugar cane and is now patented by Mitsui Chemicals in Japan (Sanghvi 2007: 21). The plasmid in the lively organisms I transformed is synthetic in origin. A translation from human language to body language: a spoken phrase, chosen from a performance that I made for *E. coli* when I started to question how this interaction might change us both. My question to the organisms is one that I hoped they would take up within the space of their bodies and repeat through many generations, much like Alvin and Mary Lucier's *Room* (1969). It is cultural, not biological information. I hoped that they would share it, I wondered if they might change it, ultimately I wanted to relate to them.

E. coli, the bacterial hosts of my thought, were once a bug in our guts, un-named until claimed by German bacteriologist and pediatrician Theodor Escherich in 1885. Now, claimed by capitalism, *E. coli* are produced and sold for laboratory research. They are popular due to their fast growth rate—described in terms of how often they double (approximately once every twenty minutes). There are thought to be more than 700 modern laboratory strains, which have evolved through treatment with agents including nitrogen, UV radiation, and X-rays as well as through simply growing them over decades. The *E. coli* family tree therefore moves from body to laboratory to shop floor. The strain I used in my work—one of the most common due to its placid plasticity—is known as "TRADE MARK: One Shot TOP10 Chemically competent E. coli." A perfectly domesticated, won't hurt a fly, do as you will with me submissive body.

I also want to draw attention to some of the liveliest materials that have impacted upon the lives of these organisms. Without the support, guidance, and good will of Volker Straub, Ana Topf, Steve Laval, and many others who have encouraged me, the lively materials of my research would not exist.

So there we have it, this lively material is all of these things and more in the summation: my thought, sugary biosubmissives, Escherich's unnamed child and all his many lab-induced descendants, all of my colleagues' time and support. I take a moment to reflect on our multiple connections. To all *E. coli bioassemblages*—past, present and future generations—my thought remains with you, for the unknowable future.

Performing Anthropomorphism

To return to the start of this essay, I have found that performing anthropomorphism brings focus to the affect of our interaction with abstracted yet lively bodies, allowing us to reconsider our actions in a relational context beyond the laboratory. Rather than anthropomorphizing the single cell, I shift focus to multiple cells, always "in this together" (Braidotti 2005). The meaning that we make through gene editing requires "an articulation of the notion of apparatuses that acknowledges [an] interrelationship of complex discursive and material practices" (Barad 2007: 199). A single cell observed through the microscope may be considered abstract. *In vitro*, the cell is always multiple and relational until reduced by technology. The same may be said for DNA. Any account of DNA in abstraction rarely conveys the liveliness of the material—its capacity to act (de Landa 2011) and its constant relation to other bodies, including our own as observer.

Following the *Transformation* workshop, participants were invited to reflect on their actions in conversation with a sentient community-being of cells from an unknowable future, the *Cells of L'Avenir*. The results form part of the short film *Zone of Inhibition* (2018), which raises ethical and ecological considerations on working with lively material in the laboratory. Accounting for our actions in the present causes pause for reflection on the experience of the other, allowing us to consider, if only for a moment, what we do in the context of the laboratory from a more-than-human perspective and what consequences this might have in a wider worldview. Bacteria and DNA exist in multitudes, both within our bodies and in our environment, and through anthropomorphizing the organism, I envisage the actions that we have in the world as transformative actions upon our extended selves. Hospitality requires us, as Derrida says, to "open up [our] home . . . to the absolute, unknown, anonymous other" (2000: 25). Through anthropomorphism, we can perhaps find a place within ourselves for this other, and thus in an aesthetics of care, I ask, *What are our obligations toward the organism, when the organism is also the self?*

Notes

1 *Oxford English Dictionary*, Oxford University Press, 2001.

2 *A Dictionary of Forensic Science*, Oxford University Press, 2012.

References

Barad, K. (2007) Meeting the Universe Halfway. Durham & London: Duke University Press.

Braidotti, R. (2005). "Affirming the Affirmative: On Nomadic Affectivity." *Rhizomes*. Rhizomes.net (11/12). http://www.rhizomes.net/issue11/braidotti.html

DeLanda, M. (2011). *Metaphysics as Ontology: Aristotle and Deleuze's Realism, European Graduate School Video Lectures*. https://www.youtube.com/watch?v=1ZjMKGTYfK4

Derrida, J. (2000). *Of Hospitality*. Stanford University Press.

Haraway, D. (1992). "The Promises of Monsters: A Regenerative Politics for Inappropriate/d Others." In Grossberg, L., Nelson, C., and Treichler, P. A. (eds.), *Cultural Studies*. Routledge, 295–337.

Karlsson, F. (2012). "Critical Anthropomorphism and Animal Ethics." *Journal of Agricultural and Environmental Ethics* 25(5): 707–20.

Lucier, A. and M. (1969). *I am Sitting in a Room* [Artwork]. https://electrocd.com/en/oeuvre/45335/alvin-lucier-mary-lucier/i-am-sitting-in-a-room

Mackenzie, L. (2016). *BioAssemblage #1* [Artwork]. https://www.loumackenzie.com/bioassemblage-1

Mackenzie, L. (2017). "Memento Perimortem [C-Type Print]." https://www.loumackenzie.com/memento-perimortem-2016-17

Mackenzie, L., and de Crécy, É. (2016). "The Genophone." Louise Mackenzie. https://www.loumackenzie.com/genophone

Nagel, T. (1974). "What Is It Like to Be a Bat?" *The Philosophical Review* 83(4): 435–50. Duke University Press on behalf of Philosophical Review Stable.

Sanghvi, Y. S. (2007). "A Roadmap to the Assembly of Synthetic DNA from Raw Materials." In Garfinkel, M. S., Endy, D., Epstein, G. L., and Friedman, R. M. (eds.), *Working Papers for Synthetic Genomics: Risks and Benefits for Science and Society*, 17–33.

Schroer, S. A. (2021). "Jakob von Uexküll: The Concept of Umwelt and its Potentials for an Anthropology Beyond the Human." *ETHNOS* 86(1): 132–52.

Servais, V. (2018). "Anthropomorphism in Human–Animal Interactions: A Pragmatist View." *Frontiers in Psychol*ogy 9 (18 December), Sec. Cognitive Science.

von Uexküll, J. (1934). *A Foray into the Worlds of Animals and Humans: With A Theory of Meaning*, trans. Joseph D. O'Neill. University of Minnesota Press (2010), 41–2, 53–5, 60–2. https://www.upress.umn.edu/book-division/books/a-foray-into-the-worlds-of-animals-and-humans. Copyright 2010 by the Regents of the University of Minnesota. Originally published in *Steifzüge durch die Umwelten von Tieren und Menschen*, copyright 1934 Verlag von Julius Springer.

8

Symbiogenesis Begins at the Mouth, Skin, and Genitalia

Ken Rinaldo

A spiked viral ball of ribonucleic acid (RNA) has made us instant kin. COVID-19, jumping from one species to the next, reveals that a "foreign" RNA can reinstruct and manipulate human DNA to produce its offspring no less. So, we nervously scrub our hands with alcohol and endure medical facemasks while fearfully darting glances at potentially infected humans. With the COVID pandemic, it is now apparent that we share cytoplasm with all others, including animals. We think of microbes and viruses as "other"; however, microbes, bacteria, fungi, and viruses are also us.

In January 2020, as the COVID-19 virus spread internationally, Art Laboratory Berlin hosted the installation *Borderless Bacteria / Colonialist Cash* (Rinaldo 2020). It consisted of international currencies soaked in liquid agar, nourishing bacteria, fungi, and viruses. Currencies sourced by Art Laboratory Berlin brought together random bacteria, fungi, and virus samples, from all over the planet. The works were produced in Germany with a protocol supplied by the artist to the Art Laboratory Berlin to curators Regine Rapp and Christian Delutz (see Chapter 12 in this volume).

Petri dishes held dueling currencies from countries at conflict with competitive microbes challenging the notion of nation-state borders. This bioart installation exposes how microbes use currency as vectors to colonize humans, all while microbes are also in conflict on our skin, mouths, stomachs, and genitalia. Still, microbial invasions of humans have been occurring since the origin of our cells, as the mitochondria within our cells are descendants of these invaders.

The endosymbiotic theories of Konstantin Mereschkowski and Lynn Margulis point to eukaryotic cells (cells with nuclei) forming our cells billions of years ago, with a symbiotic merger of non-nucleated aerobic proteobacterium. Mitochondria are organelles within living cells of nearly

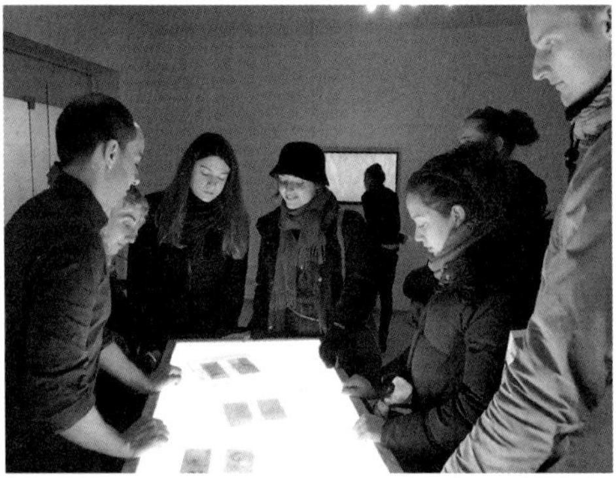

FIGURE 8.1a *Ken Rinaldo,* Borderless Bacteria / Colonialist Cash. *Installation view. Art Laboratory Berlin, January 2020. Courtesy of Regine Rapp and Christian Delutz.*

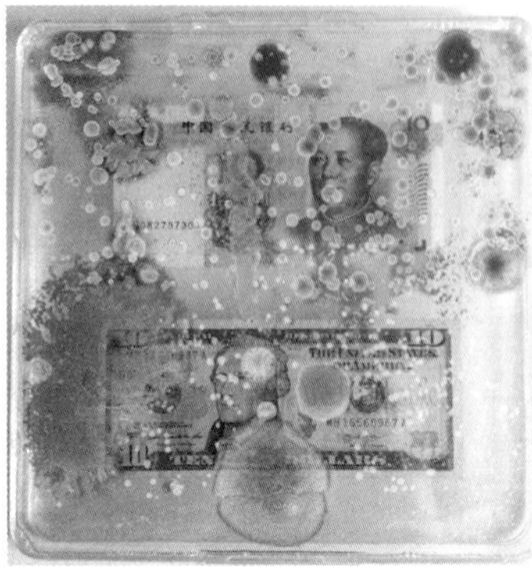

FIGURE 8.1b *Ken Rinaldo,* Borderless Bacteria / Colonialist Cash. *Installation detail. Art Laboratory Berlin, January 2020. Courtesy of Art Laboratory Berlin.*

all eukaryotic organisms (plants, animals, and fungi) which generate the chemical energy needed to power the cell's biochemical reactions. Prokaryotic organisms, such as archaea, do not have mitochondria.

The microbial colonization of human bodies happens from birth to death. It is played out with the first sip of your mother's milk entering your gut. Select parts of the human and animal body, like the mouth, nose, and genitalia, are places of entry for bacteria, fungi, and viruses. A study at NYU's Center for Genomics & Systems Biology identified 3,000 types of bacteria on dollar bills from one Manhattan bank (Bologna 2020). The microbes identified were from human skin, mouths, and genitalia; as hands are our primary method of interacting with the world, the currency is the ideal vector. Once inside our bodies, it is extraordinary that bacteria and viruses have all the keys to utilize our cells' structural nature, motor around, and reproduce.

Our ancestral microbial invaders reveal long-term structural adjustments, pathologies, and diseases resulting from being "colonized." The systemic benefits of being "colonized" may only partially arise after mutualistic relationships evolve after hundreds or thousands of generations. It is thought that our human microbiota contains about 39 trillion cells and only 30 trillion human cells.

Microbes are central to our survival, as mitochondria are the powerhouses of our cells, and viruses have also contributed to the evolution of our first cells. Viruses have shaped and built the genomes of all species, including humans. These mutualistic relationships began with parasitism leaving us with a strange core of existence—a constant battle between being a host and parasite.

By growing microbes on banknotes, we can see an implicit and explicit exchange network of mutual benefit, power relations, and challenge. This reveals interconnections of biological and economic systems and raises essential questions about the notion of "an individual self" in an age of evolving personal pronouns and identity politics. The microbiome in our stomachs further challenges the notions of the individual human self and agency, as trillions of microbes provide mutualistic benefits, tune our immune systems, and direct our behavior, all while allowing us to absorb nutrients.

The Enteric Consciousness (Rinaldo 2010) is an artificial glass stomach filled with Lactobacillus acidophilus that controls a giant robotic chair in the shape of a massive tongue. When the bacteria in the artificial glass

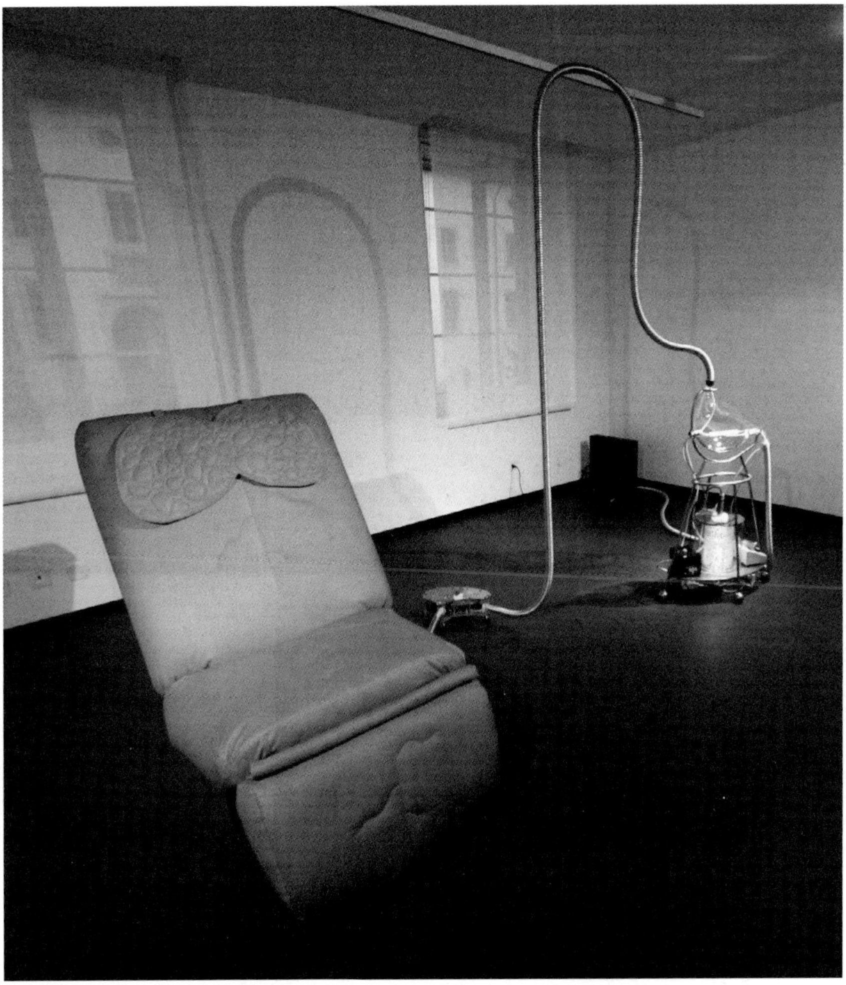

FIGURE 8.2 *Ken Rinaldo,* The Enteric Consciousness. *Maison d'Ailleurs Museum, Switzerland, 2010. © Ken Rinaldo. Photo: Joanna Avril.*

stomach is healthy, and you sit in the robotic tongue chair, you receive a 15-minute massage.

Healthy living complex bacterial cultures drive the robot's function, giving humans a corporeal experience of what it is like to be massaged by a giant robotic tongue. This bioartwork commissioned by the Maison d'Ailleurs Museum of Science Fiction, Utopia, And Extraordinary Journeys in Switzerland in 2010 functioned to help humans understand the concept of the conscious stomach. The gut (specifically the enteric nervous system) has a unique consciousness, which has coevolved with bacteria.

It is fascinating that neurons in this system interact with trillions of bacterial microbes and communicate from gut to brain via the vagus nerve and can mediate our moods. We are wet ponds of trillions of bacteria, fungi, and viruses symbiotically intertwined with human cells; we are not individuals. The artificial stomach holds the bacteria and a peristaltic pump replicating a natural gut action, moves cooling water through the bacterial cultures inside this semi-living, symbiotic bacteria/human/machine symbiont.

Two smaller red rubber robotic tongues dip into and out of chocolate in glass containers suspended with steel dopamine molecules. They draw a relationship between the evolution of our pleasure systems and bacteria that reside within us and help us process food and joy within the enteric nervous system. Enacting a corporeal experience expands interactive art, as our largest sense-organ is our skin, and there are few examples where touch is central to a work of interactive installation. The robotic tongue affects the massage experience, which is visceral and emotional.

Our enteric nervous system, which inspires this artwork, is embedded in the lining of our gastrointestinal system, and comprises one hundred million neurons, or about one thousandth of the number in the human brain. It shares the same neurotransmitters, such as dopamine, serotonin, and epinephrine, though processes experience differently in the brain. If the hand and finger can be seen as extensions of our human brain, then the tongue can be seen as an extension of the enteric nervous system, seeking out what it prefers to ingest.

While sterile at birth, the microbiome or the enteric nervous system is soon colonized by thousands of varieties of bacteria, which mediate and influence what we eat and enjoy. Childhood development points to the importance of the microbiome; as the human embryo develops, the neural crest forms and is a group of cells that evolves into both the central nervous system and the enteric nervous system. So, the brain and the stomach branch from the same group of cells. Neurotransmitters and hormones also influence the enteric nervous system, and our central nervous system can also affect stomach muscle movements, mucosal functions, and blood vessels. It is now well accepted that one of the most important jobs of bacteria in our digestive system is improving and regulating our immune systems and mediating our moods. Acid-loving milk-bacterium Lactobacillus acidophilus are the activators of the robotic tongue chair, and it occurs naturally in humans and animal gastrointestinal tracts, and after birth, the bacteria are passed from mammalian breasts to baby mouths.

The notion of the conscious stomach inspires this work and the idea that humans are not individuals so much as clouds of intertwined human, bacterial, and machine cells. We have coevolved into hybrid symbiotic ecosystems that consist of trillions of living bacteria. The bacteria within and on us are eating and surviving, and our bodies provide their sustenance. Thus, the human body is not so discrete and separate from the rest of the world.

Liping Zhao writes, "Humans are super organisms with two genomes, the genetically inherited human genome (25,000 genes) and the environmentally acquired human microbiome (over 1 million genes)." The Enteric Consciousness bio-robotic installation realizes an immersive experience celebrating this symbiotic relationship while also reflecting on a new emerging ecosystem of human-engineered robotic and web-connected entities that inhabit our homes, offices, and with nanotechnology, our biological selves.

Scatter Surge: Holobiome (Rinaldo 2020) celebrates microbial life vibrating in and on us by looking at the holobiont, a complex and dynamic ecosystem composed of hologenomes (Fester 1991). The intertwining of the host genome (us) and other associated microbial genomes determine this ecosystem's interspecies entanglements.

This installation consists of collection units (fans in Plexiglas boxes with microbe filters) that collect human skin cells along with their bacteria, fungi, and viruses. As you enter, infrared sensors and microprocessors switch seven collection units on, pulling your skin cells plus microbes inside. The samples collected are then grown on agar in Petri dishes, as holobiome-snapshots. Stones hanging in the installation reference evolutionary earth-times, and the unique mineral elements within these stones are also the substrates of life. The iron in our blood originates in dying supernova stars. Essential amino acids of our bodies have their origins in these rocks and stones, and some are fossils, where life had developed, now leaving its imprint.

On the sculptural branches of each mobile, supporting the stones and *holobiome-collection-units*, lichen is growing. Lichen is a stable symbiotic association between fungus, algae, and cyanobacteria. The origin of the term "holobiome" stems from Lynn Margulis's observations on the complex symbiotic relationships of lichens in her theory of symbiogenesis. On human skin, symbiotic microorganisms and bacteria occupy a wide range of skin niches, program vital functions that the human genome had not yet evolved. They protect against invasion by pathogenic microbes. Recent research suggests that a strain of Staphylococcus epidermidis, a common skin bacteria, can protect against cancer.

I am particularly interested in the coevolved interactions between skin cells and bacteria. That human skin cells can be transformed into pluripotent stem cells and then into any other kind of human cells, such as heart or lung

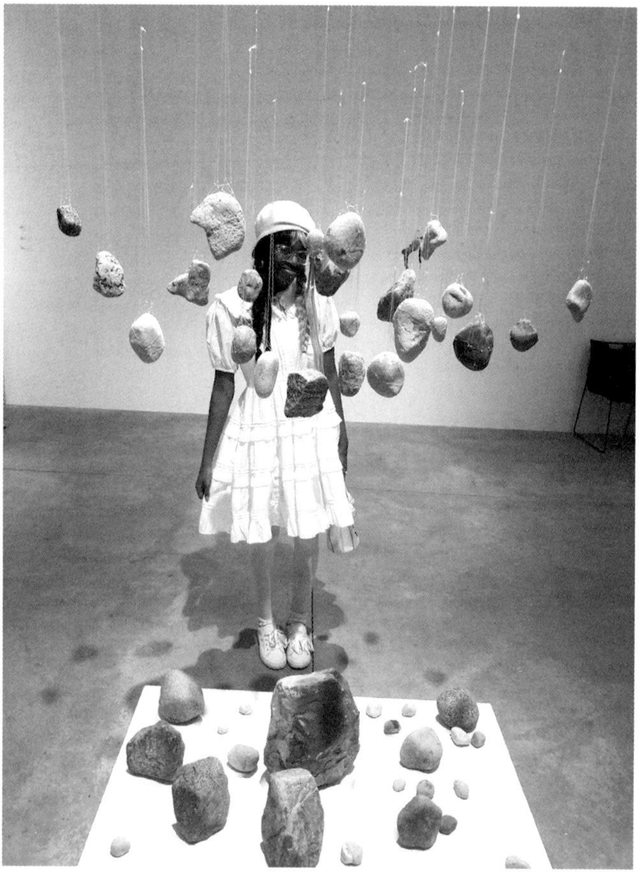

FIGURE 8.3 *Ken Rinaldo, The gravity stones within* Scatter Surge: Holobiome. *The McDonough Museum, 2020. © Ken Rinaldo.*

cells, is also interesting in this work. While the health of the body is already protected by bacteria, who would have guessed bacteria could teach us lessons about editing the DNA of viruses, leading to the discovery of the revolutionary CRISPR-Cas9 technique? Our collective (human and microbial) intelligence and mineral substrates' unique properties and material nature have also been exploited with technological advances, as silica has been refined into silicon. Computers and algorithmic logic have given rise to artificial intelligence (AI), and AI now plays a significant role in identifying, categorizing, visualizing, and evolving the natural world (ITN n.d.). AI and scientific classification have become fundamental parts of our evolution as a species. In this case, we are "becoming with" the algorithm, as they are functioning as powerful ideational amplification devices. As

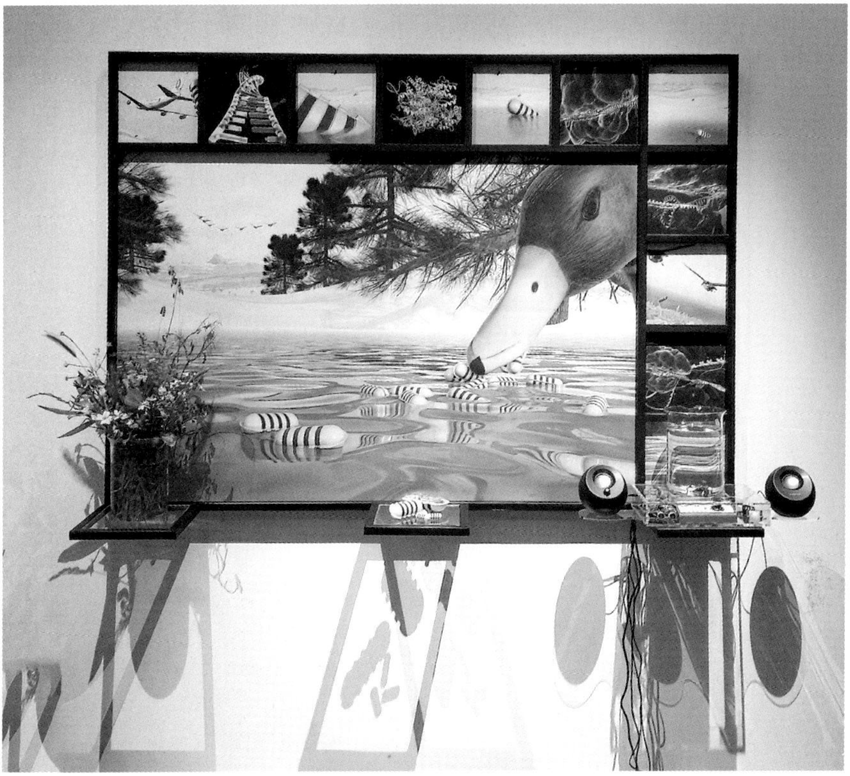

FIGURE 8.4 *Ken Rinaldo,* CRISPR Seed Resurrection. *Fundação Eugénio de Almeida in Evora, Portugal, 2021. © Ken Rinaldo. Photo: Marta De Menezes.*

silicon and robotic control join with data, we are beginning to see the potential of applying algorithmic logic to the emergence of CRISPR technologies and modifying DNA and living organisms *in vitro*. *CRISPR Seed Resurrection* (2021) is a bio artwork, that uses the evolved nature of seeds as a structural guide to creating new seeds and containers.

Part of this project presents actual seeds that have undergone CRISPR-Cas9 and X-ray and chemical mutagenic changes. They are designed to help seeds survive 10,000 years hence so that they may bloom in the distant future. The seeds entering the museum are produced by the Arabidopsis Biological Research Center at Ohio State University though they have not yet been modified to extend their sprouting time. For the CRISPR element, I am focused on Arabidopsis, as it has been genetically sequenced. Arabidopsis is a common weed related to the radish, cabbage, and cauliflower.

While the seeds entering the work have been created by the bank, I am currently researching the next phase of creating custom seeds for this work. In this first stage, Yun Wah Lam at City University in Hong Kong has specified the necessary genes to modify, to extend the future seeds' viability. The work suggests seed pods and genetically modified seeds may be distributed by planes and waterfowl into permafrost and tundra to allow a future reseeding of forest and fields. Much testing and international approvals would be necessary before releasing them, as with any new genetically modified entity, most countries have stringent laws against releasing GMO seeds as they can have unintended consequences. This work presents a high-resolution visualization of how the seeds will appear and then be distributed as a kind of social bioartwork.

While this work seems hopeful and functions as a technical and biological augmentation, it also acknowledges this may become necessary to reseed the earth, given the current rates of fossil fuel burning and global warming. Twenty to thirty percent of plant and animal species face extinction by the end of the century because of global heating (IPCC 2007). This work proposes we engineer seeds to develop immunity to higher rates of CO2 in the atmosphere and grow in future moist or dry soils, therefore extending their ability to remain viable as seeds. While most plant species reproduce with seeds, many seedless plants have become extinct, although seeds can also resurrect plants long thought to be extinct. The viability of seeds, potentially up to thousands of years, depends critically on how they are stored and kept in cold conditions.

The seeds containers I have designed allow ideal dry/moisture ratios and will maintain the seeds in containment for the longest possible time. In the imaging parts of this work, a large male mallard duck was 3D-modeled and rendered on a floating lake eating transgenic seeds that have emerged 10,000 years in the future. Mallards pick up seeds while feeding on water and land (Ross 2018). In a process called "endozoochory," some of these seeds are not digested, surviving the gut intact to sprout after passage. Mallards and other waterfowl are significant vectors for moving seeds between lakes.

Above the mallard image is a series of smaller square pictures in a film sequence taking you through time, inspired by the notion of the "cinematic metaverse" and scientific research, showing an airplane distributing seeds on glaciers. In between these are smaller 3D-modeled images of DNA being identified and modified through CRISPR-Cas9. The wisdom of bacteria-changing viruses discovered by Jennifer Doudna and Emmanuelle Charpentier has already transformed biological processes and approaches.

Its lasting impact may lie further in editing the human genome for curing disease and modifying our bodies. In the lower right-hand corner of this artwork is a beaker that violently and comically mixes miniature human figures and domestic objects in a vortex action. This activates when a passive infrared sensor detects a human nearby. It remains on for some time, implying humans are next. Miniature car parts in the vortex act to remind us of technology's promises with global heating being the disease of an oil-based economy. CRISPR is a tremendous tool that could enable the production of future seeds. However, it also empowers the modification of human DNA, introducing the possibilities of troubling future human manipulations.

The accessibility of the work is enhanced through the introduction of a song I remixed from the polyphonic music of a Portuguese Renaissance composer, Duarte Lobo, and Portuguese Fado music. A rap poem I wrote was performed by Trademark Gunderson about the tragedies of the loss of biological diversity in plants, and their seeds.

We expel blackened CO2
We saturate your air
with toxic molecules
icecaps are melting too
Acidifying—warming up
Our ocean, body steams
Our forests are burning
And so, are our dreams
We endow a sophistication
Of bacteria-fighting a virus
Genes changed, will rearrange
Allowing you, to teach us
We annotate and assemble
DNA of future seeds
That may if we need, bloom
boom and succeed
Reseeding field and Forrest
planes fly to frozen lakes
10,000 years a long time
The HONSU gene creates
We can use evolution
to change evolution
CRISPR CRISPR bloom CRISPR
CRISPR boom shine
A tug between Gibberellin
against Abscisic acid
Can we slow, down your bloom?
And create fields that shine?
Will humans witness your flower?
that ecosystems need
Will bees feel your power?
To nourish—and feed?
Will future ducks find your fruit?
and carry, from lake to lake
Drop you there, to help you sprout?
Survive and—create?
We hope these seeds will find

Warm sun, and river shore
ladybugs, and butterflies
breath oxygen to restore
a pristine place, where nature grows bacterial skills to sew
in hopes we bear witness
to the knowledge that you show
We can use evolution
to change evolution
CRISPR CRISPR bloom CRISPR
CRISPR boom shine

On a shelf in the middle of this bioart-/soundwork are actual rapid-prototyped seed pods containing CRISPR-modified seeds. To the left of the seed pods are a variety of plants and flowers in a glass beaker, showing the splendor of the natural flowering world we currently have and what may be both enhanced and potentially lost or resurrected using CRISPR-Cas9 techniques.

In the CRISPR process that I will undertake with the Arabidopsis in future versions, the genetic traits modified will be the respiratory burst oxidase homolog d, senescence-associated gene 113, *and* oresara1. The capacity to extend the life of the seed and induce seed dormancy will be added with a gene called HONSU (named after "a Korean word meaning abnormal drowsiness").

Seed germination is controlled by two hormones. Good conditions can promote the release of the hormone gibberellin, which triggers germination. Gibberellin is stopped by another hormone called abscisic acid. The genetic presence of ethylene, and the balance between gibberellin and abscisic acid, will determine the seed germination timing.

As genetically modified seeds must undergo stringent testing and implementation in controlled environments, for exhibitions, the pods and the seeds presented are rapid-prototyped samples with seeds that were CRISPR-modified by the Arabidopsis Seed Bank.

End / Begin-Again

As bacteria have battled viruses for eons, they have revealed time-evolved wisdom in the form of a genetic manipulation toolkit with far-reaching implications for humans, citizen science, and other species and how we coevolve. Gene editing happens daily by the commensal bacteria and fungi we drink in our kombucha and yogurt, as demonstrated in the Enteric Consciousness installation. Now CRISPR has empowered artists to edit genes, enabling custom bio-designed seeds.

This technology reframes questions of access, ethics, and institutional power disparities. Citizen science and CRISPR will transform our planet.

Using genome editing, we can now create phantasmagorical trans-species hybrids, with genes borrowed from other animals, microbes, and plants. Are there unseen implications that may emerge?

The CRISPR Seed Resurrection (2021) shows that seed, insect, microbe, and human genomes do not live in isolation or evolve on the same time scales as disruptive technological systems. As new genome research is further empowered by AI, algorithmic, and biological logic, the question becomes: How will gene editing impact natural environments not able to adapt quickly enough? Does the *Borderless Bacteria* and *Colonialist Cash* (2020) artwork help frame the ubiquity and benefits of bacteria, viruses, and fungi, offering ways of understanding our microbial selves and symbiogenesis? Can we come to realize our interdependence and better study, observe, and welcome the lessons they offer, so that microbes can continue to colonize, tune our health, and thrive?

The mitochondria powering our cells are a model. The ability of microorganisms to grow from an unlimited variety of food sources, coupled with CRISPR, may allow microbes of all kinds to play a critical role in helping to solve societies' current energy crisis with promises of sustainable energy. We may thus realize the ability of microbes in the soil to hold carbon and are coming to understand that microbes and healthy soils may be our best bet for saving ourselves and other species from the ravages of human-induced global heating.

References

Bologna, C. (2020). *How Dirty is Cash Anyway? HuffPost*. https://www.huffpost.com/entry/how-dirty-is-cash-coronavirus_l_5f623541c5b65fd7b8580bd7

Fester, M. (1991). *Symbiosis as a Source of Evolutionary Innovation*. MIT Press.

IPCC Synthesis Report (2007). *ScienceDaily*. https://www.sciencedaily.com/releases/2007/11/071119122043.htm

ITN (n.d.). "CRISPR Made More Precise by New Algorithm." *Informatics from Technology Networks*. https://www.technologynetworks.com/informatics/news/crispr-made-more-precise-by-new-algorithm-349758

Kaufman, R. (2021). *32,000-Year-Old Plant Brought Back to Life-Oldest Yet, Science*. https://www.nationalgeographic.com/science/article/120221-oldest-seeds-regenerated-plants-science

Ross, J. (2018). *Duck Faeces Shed Light on Plant Seed Dispersal, Journal of Ecology Blog*. https://jecologyblog.com/2018/02/06/press-release-duck-faeces-shed-light-on-plant-seed-dispersal/

9

Aphrodisiac in the Machine

Stephanie Rothenberg

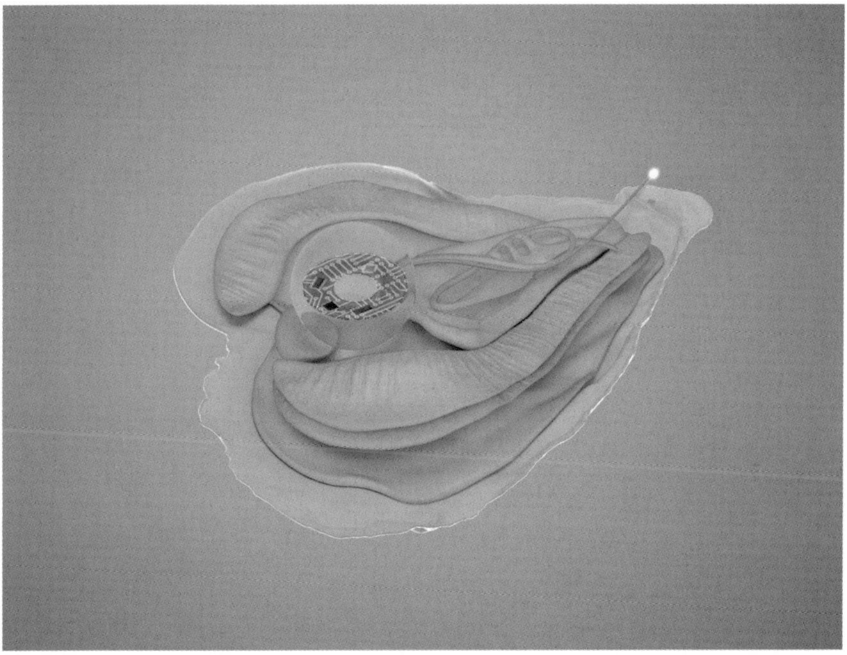

FIGURE 9.1 *Image of bioengineered oyster, 2020. Model created by Sputnik Animation. Courtesy of Stephanie Rothenberg.*

Prologue: Born from the Sea

Standing on the coastline staring at the sea, can you remember the first time you tasted an oyster?

As the smooth silky flesh slid down your throat, your body began to preprocess the molecular compound of the foreign matter. Similar to the way an oyster filters water, your thalamus, with its filtering function, known as the "gate to consciousness," was activated. This engaged your frontal lobe and hippocampus producing both a cognitive and experience-based evaluation of what was happening. Sensual. Mmmm. Top-down modulation of sensory processes began. Your previous experiences and drive kicked in causing a wave of emotion throughout your body. This triggered your amygdala driving the "fight or flight" response and causing parts of your body to react independent of your consciousness. You could hear your heart beating faster via sympathetic activation as your genitals started swelling. Your face flushed. The impulses ascended your spinal cord's spinothalamic tracts moving directly to your brain.[1] *Respiration increased drastically. You took a deep breath. And wiped the sweat from your brow. Like the goddess in Botticelli's "Birth of Venus" who was born from the foam of the sea and spawned from the parted lips of an iridescent shell, you also burst from a metaphorical shell in an utterly pleasurable state of aphrodisia. This was such a delicious moment you wished it could last forever. And maybe through a new synthesis of mind and body it can . . .*

Soft, fleshy, and resourceful, the oyster is a magnificent and extremely talented creature of the sea. It was almost extinct by the mid-twentieth century due to industrial pollution and massive overfishing. One tiny 2-inch organism can filter up to 50 gallons of polluted water per day. Its home, created from its own layered shells, combines with others to form natural reef systems that protect coastlines against rising sea levels and provide habitat for other species to thrive. And legends speak wonders of its euphoric powers as an aphrodisiac.

Imagine if we could bioengineer this magical species to transform toxic water into an even more transformative formula and pipe it into public drinking water? Could we create a sentient machine of more perceptible humans? A perception that enables a more sensual interconnection with the cycle of life that leads to a better handling of this ecological crisis?

Aphrodisiac in the Machine is an environmental science fiction that manifests in a variety of formats online and offline including art installations, videos, and performances. Through the lens of a genetic manipulation intended to thwart the devastating effects of climate change, this artistic project explores the ethical and techno-economic contradictions within the desire to be more sustainable. The neoliberal concept of natural capital and its unfolding through what is known as ecosystem services is repackaged, illuminating the constant bioengineering of living systems required to maintain the economic agendas of these supposedly sustainable models.

In the narrative, the libidinous myth of the oyster, a hermaphroditic organism, is center stage as the oysters are bioengineered using the popular gene editing tool CRISPR to create a viral mutagenesis. The creatures are grown in futuristic aquaculture farms, a form of sea-based farming, located near public drinking water systems. Through this genetic manipulation, the desire to design the perfect organism subservient to human needs is given a twist. Biotechnology becomes eroticized as the genetically modified oysters convert polluted waters into an aphrodisia-inducing water that is freely accessible to drink from any faucet. By ingesting the water, it is the human that is now being redesigned.

In *Aphrodisiac in the Machine,* the erotic is what black feminist writer Audre Lorde calls a power of feeling. In her essay, *The Uses of the Erotic, The Erotic as Power*, published in 1984 (Lorde 1984), she challenges the patriarchal overtones in how the word "erotic" is used, not only redefining but reigniting the erotic as a physical, psychic, and emotional energy that can't be reduced to a commodified good or systematized affect. Using Lorde's notion of the erotic, *Aphrodisiac in the Machine* posits more-than-human sentience as a lubricant to speculate a new kind of eco-machine. By asking if this new and improved bioengineered oyster can push humans past the mere libidinal and sexualized state of capital conquests of other bodies and into a new state of sentience—a Sentience 2.0—the project raises critical questions about the use of gene editing tools such as CRISPR. Can we use gene editing tools to shift human and other-than-human relationship models to be more symbiotic? And with new citizen science platforms, where does the power of creation and governance reside?

In this experimental chapter, two juxtaposed perspectives move between the speculative and the factual. Each section begins with the corporate voice of Aquadisia Networks (in italics), a fake biotech company that is bioengineering a new species of oyster to "cure" climate change through a supposedly medicinal water. The speculative voice is followed by research (non-italicized) in the areas of marine conservation, aquaculture, and gene manipulation obtained from various sources including scientific articles, industry publications, and cultural magazines. This content serves as a departure point for creating the project narrative. By pushing the factual into a fantastical and rather absurd storyline, the project aims to not only critique some of the contradictions of blue capitalism but provide space for envisioning alternative multispecies future-making.

Aquadisia Networks

Hovering above the crashing waves of the high-tech industrial coastline of Hawaii's "Silicon Valley" is the home base of Aquadisia Networks. We've been working hard to earn our reputation on what we know best, life below the water. After spending years attempting to develop sustainable solutions for the global sea farming industry and only running into dead ends, our technologists had an epiphany. The link between conservation initiatives

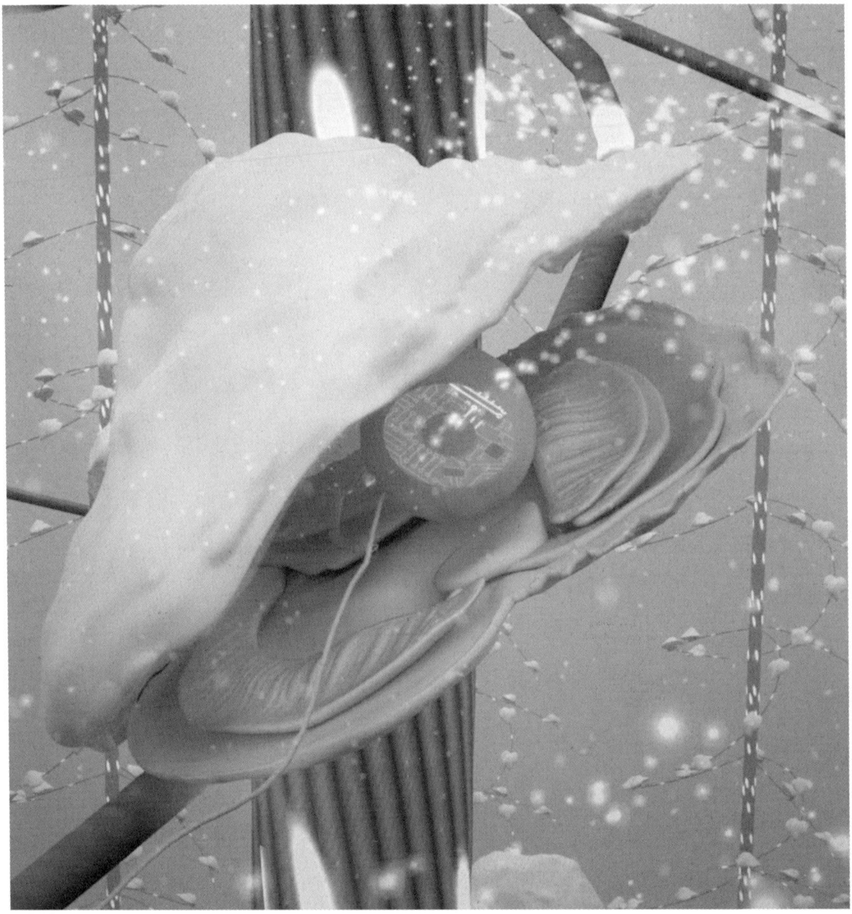

FIGURE 9.2 *Image of Aquadisia Networks aquaculture system, 2020. Model by Sputnik Animation. Courtesy of Stephanie Rothenberg.*

and food security was broken. These so-called value-added solutions that claimed to put "the environment at the heart of the economy" were just new wrappers of green capitalism for the "blue economy." In order to truly cure climate change a revolutionary new model was needed—a model that could confront our anthropocentric crisis head on.

At Aquadisia Networks, we believe there is only one way to do this—reinvent the environment's nemesis, the human. This radical way forward requires the rethinking and reengineering of three key components: natural capital, aquaculture, and the Mollusca ostreidae (the magical oyster). Together this triadic combination creates a powerful new vision for future survival.

Unnaturally Flavored Capital: Ecosystem Services

In business schools there is a common saying: "What you do not measure, you do not manage." This is the philosophy of what is known as natural capital—managing the earth's ecosystems to meet the demands of the market. But there seems to be something fundamentally wrong with treating the earth as if it were a business undergoing liquidation. There is nothing natural about natural capital. Tuning nature to the rhythms of the economy is like training a fish to invest in the blockchain.

Aquadisia Networks believes another world is possible. It offers a beyond-the-human model that doesn't view nature as separate from humans. A model that doesn't just integrate the capitalist into the web of life, but transforms them. This isn't a new idea. We just didn't have the technology until now. Advancements in biosystems engineering are enabling a new reality that is making natural capital actually natural.

We've redesigned the current human-centered "ecosystems services" model with a revolutionary new approach to thinking about sustainability—an "ecosystems hospitality." In Figure 9.3, the diagram on the left depicts the current model that puts humans in the center as utilizers. In reengineering that model, the diagram on the right now envisions the center to encompass an all-species approach to providing but also receiving. Cartesian dualism no more, through a symbiotic mutualism the earth's "technologies of enchantment" can prosper.[2]

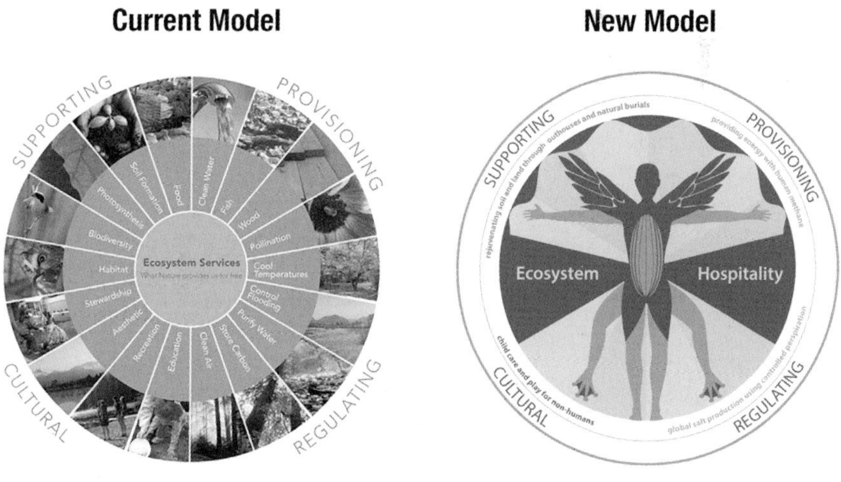

FIGURE 9.3 *The new model on the right provides just a few examples of an "ecosystem hospitality." Left image source: IUCN/International Union for Conservation of Nature. Right image courtesy of Stephanie Rothenberg, 2021.*

Rendering nature visible to the demands of capital is enacted through the concept of ecosystem services. These are categorized into services that provision and regulate nature's "goods," provide support systems that aid reproduction such as nutrient cycling and biological habits, and lastly provide cultural services. All are designed to maintain biodiversity within capitalism's need for extraction through market-based governmental policies.

But there are many contradictions and ethical concerns in what British environmental anthropologist Sian Sullivan calls the "green growth development paradigm" (2017: 397). Sullivan questions what is at stake when nature is instrumentalized as a "set of assets" that "can be valued in economic calculations" (397). Critiquing the work of the UK's Natural Capital Committee, an independent advisory committee formed in 2012 to advise the UK government "on the sustainable use of natural capital," she underscores the "diversities [that] are lost in the world-making mission to fashion and fabricate the entire planet as an abstracted plane of (ac) countable, monetizable and potentially substitutable natural capital" (398).

Scientists and environmentalists have been researching ways to protect and sustain biodiverse ecosystems and the services they provide within the sphere of nonhuman life since the turn of the century. The shift to a quantifiable, commodity-framing of nature, to provide human services, became more widespread through initiatives that brought together industry executives with global policymakers alongside "career environmentalists." "The Limits to Growth (LTG)", a report published in 1972, was one of those turning points. The report presented a post-Malthuisan analysis[3] that was the first to use a computer simulation to illustrate the finite supply of resources in relation to population growth. Although the report was highly criticized, it helped establish the framework and rhetoric around the concept of "sustainable growth." Almost thirty years later in 2000, the United Nations Millennium Ecosystem Assessment became a global acknowledgment of the anthropogenic crisis and was instrumental in inserting the terms "natural capital" and "ecosystem services" into the neoliberal vocabulary of current mainstream conservation policies.

Aquaculture: Forever Oceans and Giant Toilets

At Aquadisia Networks, we are making a difference using a truly eco-friendly innovation that will revolutionize sea farming forever. Our unique offshore aquaculture vessels are being deployed in polluted waters around the world. These fully automated systems are home to a new breed of the cyborg Mollusca ostreidae that filter toxic water and transform it into refreshing Aquadisia Water.

Our bioengineered organisms are grown in situ using a remote command and control interface that enables sea robots to conduct basic organism husbandry in the vessels. Biosystem technologies, using cloud-based software, streamline production and detect risks. Biometric sensors monitor organism health around the clock ensuring precise quantities of fluid output. Aquaculture vessels are located in proximity to municipal water sources to allow a continuous flow of secretions into public drinking water. Access to Aquadisia Water is as simple as turning on your home or office faucet. And if you don't like oysters, no problem. Aquadisia Water tastes like water.

Aquaculture has a long history evident in ancient China, Egypt, and Rome as well as Mayan societies and North American native communities, but this last century has experienced rapid innovation. Massive decline in fish stocks from overfishing and pollution led to development of marine laboratories and fish hatcheries. Hormone manipulation, food pellets, floating cages, and the advent of rust and corrosive free plastics revolutionized the industry. Modern pisciculture is on the rise, making "an acre of water more valuable than an acre of land" (Isabella et al. 2020).

As higher production costs and environmental politics around deforestation and factory-farmed pollution continue to challenge land-based agribusiness, industry has turned to aquaculture as a source of protein-based food solutions. From 1990 to 2018 there was a 527-percent increase in global aquaculture production and 122-percent increase in total fish consumption (Aquaculture Alliance 2021). Recent advancements in biosystems engineering elevate aquaculture to a new level. What began in the 1990s as acoustic telemetry, in which fish were fitted with miniature ultrasonic transmitters to quantify their behavior, has moved into the realm of highly sophisticated biometrics and genetic engineering. Genetic manipulation creates more resilient species while computer vision conjoined with artificial intelligence monitors the quantity, movement, and health of individual fish within a school of millions.[4]

Key to aquaculture's marketing strategy has been the buzz word "sustainable" alongside the threat of expanding human populations—which is estimated to be 10 billion by 2050 (Suzuki 2019)—to position it as an eco-friendly alternative. Fisheries ecologists at the University of California Santa Barbara estimated the amount of ocean available for aquaculture to be over 4.6 million square miles (Borrell 2020). This is enough water mass to produce 100 times the amount of seafood currently being consumed. These same ecologists admit that these initiatives may put many conservation areas in direct conflict with aquaculture's economic goals. Large-scale aquaculture ventures such as the salmon farming industry have already "transformed some of the most sensitive and valuable aquatic ecosystems on Earth into a giant toilet" (Borrell 2020).

Aquadisia Water: The Science of Sentience 2.0

Key to this revolutionary new model is the Mollusca ostreidae—scientific term for the oyster, and what we view to be a magical creature. Through its reengineering, humans will now be able to achieve an experience that goes beyond the libidinal limits of aphrodisia. We call it Sentience 2.0, expanding the realm of normal human perception and mere sexual arousal into a new energetic awakeness that eroticizes sensation into beyond-the-human sensorial ability. Combined with our other two reengineered components—aquaculture and ecosystem services—we believe the Aquadisia experience can cure climate change.

In the laboratories at Aquadisia Networks, technologists are developing a new breed of the Mollusca ostreidae that leverages the organism's ability to produce D-aspartic acid, the essential ingredient to aphrodisia. Using the gene editing program CRISPR, the organism is genetically altered to produce an advanced formula of D-aspartic acid. When ingested by humans it will result in enhanced activation of a wide range of neuroendocrine cells in the body that have the potential to induce a state of Sentience 2.0.

Human experience shifts from insatiable consumptive desire to a more sensorial and energetic pleasure that can be derived from simple, everyday experiences like bathing in a mountain stream or rejuvenating one's yard. With a new sense of awareness toward both humans and nonhumans, our actual needs and their impact on the environment can be reevaluated. This new state of being creates a trickle-down effect as industry and government leaders who are drinking Aquadisia Water follow this trajectory, changing the reality of what a sustainable economy is—an economy that provides for society without the production of unnecessary waste and excessive extraction. At Aquadisia Networks, we see this as a global shift involving all stakeholders with the goal of achieving a herd Sentience 2.0 for future survival of all species.

In 2005, a scientific study was conducted by a chemistry professor and his colleagues at Barry University in Miami, Florida, to determine if oysters were actually an aphrodisiac. The team discovered that oysters (and the family of bi-valve mollusks) produced a high level of the amino acids D-aspartic acid and N-methyl D-aspartic acid (NMDA) which are associated with increased sexual hormone levels in human and nonhuman species.[5] When injected into laboratory rats they effectively demonstrated their sexual prowess.

Although the study was inconclusive on the organism's potential as an aphrodisiac in humans, it gained widespread attention, circulating throughout mainstream online sources even ten years later. Articles with titles such as "The Weird History of Oysters as Aphrodisiacs" in well-known media outlets such as WIRED and Smithsonian Magazine helped perpetuate the myth of Casanova and his unstoppable libido. As noted in the eighteenth-century Venetian explorer's explicit sexual memoirs, he consumed fifty oysters for breakfast every day (Casanova [1768] 2001). What has been somewhat more conclusive and popular in the scientific research of these amino acids is the use of D-aspartic acid as a testosterone booster for infertile men and by athletes to increase muscle mass (Furst et al. 2017). When ingested the amino acid affects the production of growth hormones leading to its popularity as a supplement for sports enthusiasts.

D-Aspartic acid is an endogenous (internal) amino acid occurring in specific tissues and cells of both invertebrates and vertebrates. It is found in high concentrations in the human pineal gland, that special spot located between the two hemispheres of the brain which French philosopher René Descartes called "the seat of the soul." It is in this sweet spot of the so-called soul that Audre Lorde's theory about the power of the erotic reemerges.

For Lorde the erotic is not sexual nor is it pornographic which she explicitly states is sensation without feeling. The erotic is an untapped resource "rooted in the power" of unexpressed feeling that resides within us from a "deeply female and spiritual plane" (1984: 53). It is the energy and love we want to be able to put forth into both our everyday experiences and higher-reaching goals, feeling satisfaction, not mediocrity. Through Lorde's feminist lens, this power of feeling must be reclaimed in order to move beyond dominant patriarchal power structures. And over thirty years later from when her seminal text was written, these same uneven power structures are the forces perpetuating our ecological crisis.

Conclusion: The Viral Aphrodisiac in the Machine

The metaphors of "man as machine" and the desire to humanize the machine, to make it sentient and aware of feelings and sensations, has a long history. René Descartes's signature doctrine of mind-body dualism was instrumental in creating the Western philosophy of mechanized bodies and disembodied souls. The function of atoms, cells, and organisms are continually made understandable through Cartesian machine analogies.

But with advancements in emergent machinic behavior and bioengineering, is the simplistic dualism of life vs. machine finally breaking down? The ability to integrate machines with living systems—from body parts down to neurons and even viruses, are creating new global biological minds. *Aphrodisiac in the Machine* addresses and amplifies these scales specifically within the context of viral culture. Microscopic genetic invasions conducted on the oyster's physiology result in a larger cultural "viral" shift as the public is transformed by drinking the bioengineered *Aquadisia Water*.

In economic theory, machines are designed by humans, but it is also the "invisible hand" of the market that enables certain machines to proliferate. This creates an interesting analogy between human provenance and evolutionary processes (Bongard et al. 2021). In relation to machine metaphors, I bring bring forth one of my favorites—"to throw a monkey wrench in the machine." Through the lens of Audre Lorde's notion of erotic power, *Aphrodisiac in the Machine* speculates on sensation as the wrench to disrupt and destabilize the hegemonic and oppressive narrative of blue capitalism. It becomes a tool to reclaim power through sentient experiences and actions to establish a deeper sensual connection with the cycle of life. *Aquadisia Water* becomes a hypothetical medicinal "cure" to the threat of climate change and shifts the oyster myth to a source of erotic empowerment vs. sexual subjugation.

There are already plenty of ideas for future machines. But we still can't figure out how to intervene in the exploitative, systemic machines of capitalism. Perhaps then, we must ask, what is the wrench? In this story it's an aphrodisiac like no other that transforms human sentience into an agential erotic power. Hopefully it would make Audre Lorde smile.

Notes

1 The description of the physiological effect of sexual arousal was combined from articles by Roy J. Levin and Andreas Baranowski/Rudolf Stark.

2 This idea is taken from Alfred Gell's concept of the "Technologies of Enchantment." Art is created through the use of technology whether a traditional wood-carving implement or digital platform. Through these technologies the artist is able to depict reality in an enchanted state. In the essay, this enchanted state refers to the power and reverence of nature.

3 "An Essay on the Principle of Population" written in 1798 by economist Thomas Robert Malthus.

4 Major innovators in the aquaculture industry are based in Hawaii's Silicon Valley of aquaculture.

5 The poster session was presented by marine scientist George Fisher, formerly at Barry University in Miami, and his colleagues in 2005 at the American Chemical Society.

References

Aquaculture Alliance, https://www.aquaculturealliance.org/education/

Baranowski, A., and Stark, R. (2019). "Sexual Differences in the Perception of Sexual Arousal." In Sesko, A., and Baldwin, M. (eds.), *The Inquisitive Mind Magazine*, Issue 42: Gender Special Issue, December. https://www.in-mind.org/article/sex-differences-in-the-perception-of-sexual-arousal

Bongard, J., and Levin, M. (2021). "Living Things Are Not (20th Century) Machines: Updating Mechanism Metaphors in Light of the Modern Science of Machine Behavior." *Frontiers In Ecology and Evolution* 9. https://doi.org/10.3389/fevo.2021.650726

Borrell, B. (2020). "Show Me the Money Fish." *Hakai Magazine*, August 24. https://www.hakaimagazine.com/features/show-me-the-money-fish/

Casanova, G. ([1768] 2001). *The Story of My Life*, Pizzamiglio, G. (ed.). Penguin Classics.

Fisher, G. H., Mirza, R., Poisson, J.-P., D'Aniello, A., Spinelli, P., and Ferrandino, G. (2005). "Do marine mollusks possess aphrodisiacal properties?" American Chemical Society, Poster Session: MEDI 401, March 16. http://oysterhealth.com/hcvnews/7.html

Forever Oceans Corporate Website, "Who We Are." https://www.foreveroceans.com/who-we-are/

Furst, T., Horvath, P., LaMacchia, Z., William, B. (2017). "Acute D-Aspartic Acid Supplementation does not have an Effect on Serum Testosterone but does have an Effect on Strength Measures in College Aged Male Athletes." *European Journal of Sports & Exercise Science* 5(3): 34–41.

Gell, A. (1994). "The Technology of Enchantment and the Enchantment of Technology." In Coote, J. (ed.), *Anthropology, Art, and Aesthetics*. Clarendon Press.

Isabella, J., and Hunt, S., (2020). "A Short History of Aquaculture Innovation." *Hakai Magazine*, August 24. https://www.hakaimagazine.com/features/a-short-history-of-aquaculture-innovation/

Karimi, J. (2017). "A Step to Sustainability; MAES Mapping and Assessment of Ecosystem Services in European cities and Italy." *IUCN*. https://www.iucn.org/news/commission-environmental-economic-and-social-policy/201702/step-sustainability-maes-mapping-and-assessment-ecosystem-services-european-cities-and-italy

Levin, R. J., (2002). "The Physiology of Sexual Arousal in the Human Female: A Recreational and Procreational Synthesis." *Archives of Sexual Behavior* 31(5): 405–11.

Lorde, A. ([1984] 2007). *Sister Outsider, Essays and Speeches by Audre Lorde*, "Uses of the Erotic: The Erotic as Power." Crossing Press, 53–9.

Malthus, T. R. (1798). "An Essay on the Principle of Population." History of Economic Thought Books, McMaster University Archive for the History of Economic Thought. https://historyofeconomicthought.mcmaster.ca/malthus/popu.txt

Marx, K. (1887). *Capital: A Critique of Political Economy*, Engels, F. (ed.), 1:163. www.marxists.org/archive/marx/works/download/pdf/Capital-Volume-I.pdf

Roshanzamir, F., and Safavi, S. (2017). "The putative effects of D-Aspartic acid on blood testosterone levels: A systematic review." *International Journal of Reproductive Biomedicine* 15(1): 1–10.

Sullivan, S. (2017). "Natural Capital, Fairytales and Ideology." Invited Review Essay, *Development and Change* 48(2): 397–423.

Suzuki, E. (2019). "World's population will continue to grow and will reach nearly 10 billion by 2050." *World Bank Blogs*. https://blogs.worldbank.org/opendata/worlds-population-will-continue-grow-and-will-reach-nearly-10-billion-2050

PART THREE

COVID and the Arts: Reflections

10

Art Worlds Evolving

Notes on Evolutionary Metaphors of Change and the Global Art System

Meredith Tromble

As I write, my local art world in Northern California is cautiously reopening from a year of pandemic isolation, the year of the COVID-19 virus. You, the reader, reading in your own, future time, will not have experienced this turbulent year in exactly the same way, or perhaps even experienced it at all. What is important for you to know about this writer's standpoint, her view into change in the global art system?

This writing began with an artwork, in the ideas bubbling around my artist residency at the Complexity Sciences Center at the University of California, Davis, where I made an interactive virtual reality work called *Dream Vortex*. Looking for an image that embodied complexity, I photographed boiling water, a classic example of turbulence. Before we turn from artwork to essay, I invite you to picture along with me. Imagine a vast stretch of boiling water, extending to the horizon in all directions. Dive into this image with me: pretend, for a moment, that this is our world. We can stand on a bubble as if it were a stepping stone, or surf the ripples.

But there's no steadiness, no predictability, everything is moving. This is what happens when you pour heat into a system: flux. This image of a world in flux is both the emotional backdrop for my discussion, and a metaphor for the pandemic's ambient reality of carbon-induced climate instability. Heating water drives a state change from liquid to gas, from water to steam. Perhaps state changes in the art world—in the institutions and relationships

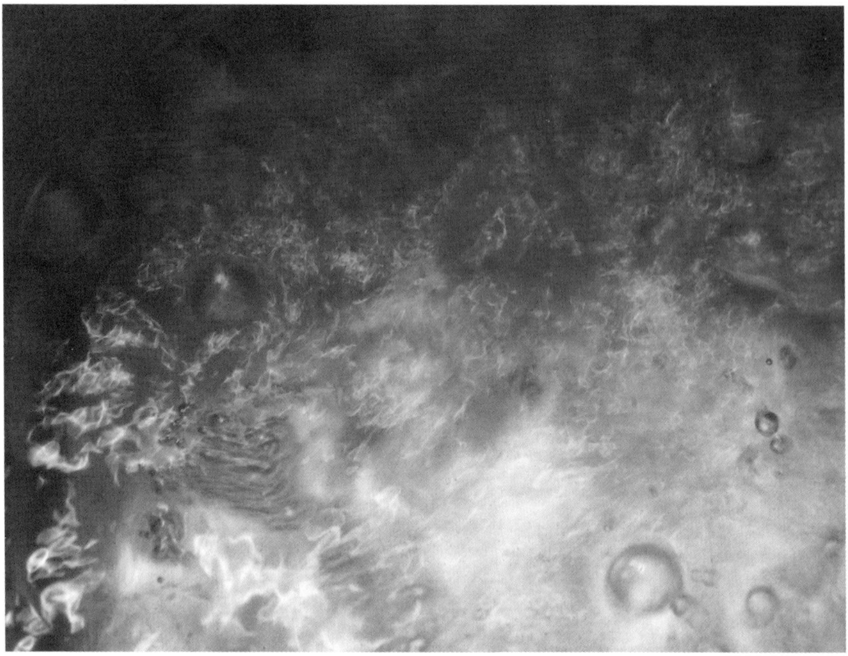

FIGURE 10.1 *Meredith Tromble and Dawn Sumner. Detail of boiling water "landscape,"* Dream Vortex, *2011–19. Courtesy of the artists.*

that support the production of contemporary art—were set in motion by the pandemic. Given the likely connection between the pandemic and climate change (Lorentzen et al. 2020), there is the reasonable expectation of continued post-pandemic turbulence in the art world, as in the social systems that produce and maintain it.

Without attempting to predict the future, which seems a fool's endeavor amid these changing conditions, I hope to capture some meaningful questions that you, too, might have . . . you, a reader who cares enough about contemporary art to read about its "evolution," and thus by definition a member of the global art system. You, a node in the great elastic network of imagining, making, marketing, and receiving art, whatever art might be. What meaning are we, as art world makers, to make of this time. How are we to think of the art world, post-pandemic?

To start, we may notice how we are thinking about change, now. How are we describing the shock to our system?

Magda Sawon, co-founder, Postmasters Gallery:

> The current virus may well be the equivalent of the asteroid that killed the dinosaurs. Because sometimes, the system has to be destroyed in order to be liberated—and to make room for evolution. (Sawon 2020)

Mitchell Algus, art dealer:

> . . . the art world, like many other communities, is passing through an evolutionary bottleneck and its genetic structure is about to change in unforeseen ways. Which organisms—galleries, that is—will squeeze through to the other side is as yet undetermined, as is what they will look like when this ends. But change will happen. (Algus 2020)

Jerry Saltz, art critic:

> Over the last decade or so, the art world in peril has seemed to lose the ability to adapt. (Saltz 2020)

Sawon, Algus, and Saltz use evolutionary language and metaphors to talk about change in the art system. They speak of destruction, unpredictability, and peril along with change and opportunity. How do these evolutionary metaphors make you feel? Do they make you want to dive under a rock? Push your way to the bottleneck? Head for the exits? I think I felt all of these things in the past year. When there was serious talk of trusting citizen's lives to "herd immunity," underpinned by a seldom-spoken assumption that "the fittest would survive," I felt despair. But sensing that fear and anger were shuttering my vision, I began looking for openings. Evolutionary metaphors, which speak to the entanglement of living beings with their environment (Brooks 2011), have often been pressed into the service of a nasty politics of fear and dominance of the fittest. But that is not the only way they can be used. The study of evolution has many useful things to say about our snarl of natural, social, and economic catastrophes. It is one of the most relevant, encompassing bodies of knowledge we can draw upon.

There are other metaphors we could use to think about the art world. The critic Jack Burnham used the metaphor of "information," writing: "A major illusion of the art system is that art resides in specific objects . . . But in essence, all institutions which process art data, thus making information, are components of the work of art" (1969: 30).

In a moment I will discuss a metaphor that integrates information with evolution, but first let me reveal why I care about the action of these metaphors. Let us consider how they work in relation to the question of how should I, how should we—as artists, audiences, and art world makers participating in the global art system—respond to our changing circumstances?

A Case Study

Early in March 2020, the San Francisco Art Institute—the school where I am a professor—"suspended academic programs indefinitely . . . as the result of a long-brewing financial crisis that was disastrously worsened by the pandemic," as described by Brian Boucher in *Artnews* (2020).

On March 17, 2020, the City of San Francisco implemented a "shelter in place" order for all residents (Allday 2020). The last students left the building. The guard made rounds to lock the doors. The bicycle rack was down to bikes left by students stranded off-campus. The parking lot emptied.

A few days later, all faculty and staff received layoff notices. It was grim. There were endless official and unofficial conversations about what to do. I heard talk of "competition" and "predatory lenders." I heard that there was a "struggle for survival" in our "institutional environment." I began to wonder how this language interacted with our collective understanding of our options and potentials. By June it was clear that these problems were not ours alone. We had a particularly bad case of pandemic woes, but higher education in general thrashed through changing in-person classes to online instruction, changing a predictable schedule to "nobody knows," changing from international circulation to study at home. We went from a world where planning could occur on the basis of historical information to a world in which the past had a highly unpredictable relationship with the future.

In response to a survey on the impact of COVID-19 on higher education conducted by the Center for Organizational Leadership at Rutgers University, an anonymous dean wrote, "My analogy here is the punctuated equilibrium theory of evolution. There are times or events that turn the world on its head . . . Unfortunately, those periods are ones in which there is loss, and I believe that we will see a great deal of 'radiation' in higher ed, but also a lot of 'extinction'" (Gigliotti 2020).

Yet a year later, I am here writing, still a professor, still an artist with an institution, and a teacher with students. Love for the school and hard work brought aid from others—government, alumni, donors—that lets the school exist now (Boucher 2020).

In this past year-without-a-future, the school turned inside out. The people left the building, but we kept making the school. An outdoor exhibition program kept the building alive night and day, projecting art films on the school's tower and showing a huge sculpture by Rigo 23 of Indigenous activist Leonard Peltier in the school plaza, facing Alcatraz Island (the site of a historic Native American occupation claiming the land for all the tribes of North America). We taught online while the school community met virtually to "reimagine" the school. Now we find ourselves one step beyond imagining a future, bringing those imaginings into conversation with the world as it exists.

Past Present

The world as it exists is the product of the past, which has given us imagery and turns of speech that have become "naturalized." But evolutionary metaphors are young enough that their history can be traced with some surety. Charles Darwin's *On the Origin of Species* was first published in 1859; in 1888, the Canadian-English journalist Grant Allen wrote, "Everybody nowadays talks about evolution" (Clarke 2010: 112). Discussing the results of that talk, the absorption of Darwinian concepts into literature, from whence they traveled into contemporary vernacular English, scholar David Amigoni writes: "Everybody continues to talk about evolution . . . Charles Darwin's manipulation of the written word—his willingness to fashion imaginative analogies and metaphors to advance his case is unquestionably more compelling and involving for the readers . . . multiple meanings of and relations to the 'literary' have been fashioned in the forging of evolutionary theory" (2010: 113).

The three uses of evolutionary metaphor that I cited above are linguistic relics, messengers from the Victorian context in which the science of evolution took shape. In that context, a god that directed all of existence was assumed. Behind the now-vernacular language drawn upon by Magda Sawon, for example, is an unacknowledged entity filled with purpose and desire. When she says, "sometimes, the system has to be destroyed in order to be liberated—and to make room for evolution," a god that desires evolution—an external power for which evolution is an imperative—is implied in the language.

When Mitchell Algus applies the metaphor of genetic structure—conceived of as a template by which ongoing life is constructed—to describe the cultural commercial forms that produce art galleries, he participates in a long tradition of applying biological concepts about living bodies to human social worlds. Early in the historical discussion, evolutionary concepts about the "natural" realm were extended to describe the human cultural realm (Beer 2009: 2). Darwin's language about "natural selection" was combined by his contemporaries with Herbert Spencer's term "survival of the fittest" (Darwin Correspondence Project 2020) in eugenic discourse about "improving the genetic composition of a population through controlling reproduction" (Subramaniam 2014: 46) as well as framing scientific discourse about variation in living bodies.

This is not to say that Sawon or Algus consciously derived the evolutionary metaphors they used or to imply that their individual thinking is Victorian; rather I make the point that the language of evolution as developed in the nineteenth and early twentieth centuries is embedded in everyday American speech (not to mention decorating quite a few American t-shirts and bumper stickers). It would be easy to overlook how strange a comment such as Saltz's, that assumes "the art world" could once do something called "adapting," might seem in a cultural context where evolutionary science had never happened. It is also common to overlook the ways in which the contemporary science of evolution is actively rejecting, elaborating, and transforming Victorian ideas about change in life forms.

Just as Victorian evolutionary science was colored by Victorian culture and politics, contemporary evolutionary science is vibrantly and complexly connected with twenty-first-century society. For example, it was a female scientist, Lynn Margulis, who formalized the concept of symbiosis as a key evolutionary force, and fought for this counternarrative of cooperation, rather than competition, through decades of rejection until, finally, evidence largely brought the field around to her way of thinking (Khalil 2012: 19). In contemporary culture, vital insights into the ways "the institution of science reified social power and privilege into claims of natural and biological difference" (Subramanian 2014: 12) challenge and coexist with patterns of thought and practice still shaped by Victorian mores, just as contemporary evolutionary research encompasses both symbiosis and selection. The multiple theories that together form our contemporary understanding of evolution generate more models and metaphors than simply "survival of the fittest."

Bet-hedging, for example. I'll start with a quick example of "bet-hedging" in a scientific context, referring to those ancient survival experts, bacteria.

In 2018, physicists Sarah Marzen and Jim Crutchfield published a paper describing the epigenome—the factors around the genome that determine how it will be interpreted—as "a mechanism by which organisms can remember the environmental past" (Marzen 2018: 1). Here is what they say in the first lines of their paper, paraphrased in lay language: Individuals within a population of bacteria can have exactly the same genes but different bodies (phenotypes), because there are always imperfections in how the genes are reproduced and translated. The variety this "noise" produces in bacterial bodies gives evolution something to work with, but instead of transitioning to a "fittest" state, as selection acts on a population, that population may hold on to phenotypic differences, increasing its capacity to thrive in more than one environment. This is "bet-hedging."

Difference Matters!

Marzen and Crutchfield continue: "[I]t may seem strange that a population of organisms should not simply express the phenotype that grows best in the most probable environment . . . Imagine, however, that the environment fluctuates somewhat unpredictably . . . If organisms only express that single phenotype, then, eventually, the population will go extinct . . . Epigenetics provides a mechanism by which organisms can remember the environmental past," keeping more options open for the organism (2018: 1).

Marzen and Crutchfield develop an argument about the value of epigenetic information for populations subject to fluctuating environments in mathematical terms. They write that it "suggests evolution favors populations of organisms that develop epigenetic memories which [recall] as much of the environmental past as possible." Using terms from information theory, they describe epigenetic memory as "side information about future environmental states," which increases the chance that a population of organisms will survive (2018: 1).

Crutchfield directs the Complexity Sciences Center where I was in residence, and so I read their paper. Their description of "side information" that became useful in "conditions of flux" caught my imagination. Might minding side information identify new courses of action for art world makers? Were there art world equivalents of "bet-hedging" and "epigenetic memory"? Could bet-hedging help our global art world survive? And would this rather offbeat metaphor be more productive than already available dictums such as "don't throw the baby out with the bathwater"? Yves Salomon-Fernandez, a college president, mixed that homely metaphor with evolutionary imagery when she said: "Our sector is married to tradition and, at times, we appear unable to disentangle ourselves even if the cost is our own extinction. This does not mean throwing the baby out with the bathwater. In a changing world, how do we learn to preserve the culture and characteristics responsible for our success, thus far, while making room for experimentation, failure, success and evolution?" (2020).

Or could thinking about "bet-hedging" help us more than the straightforward acceptance of unpredictability? That could also be a stance, as stated so beautifully by artist Shilpa Gupta, who said: "Perhaps we are searching for predictions in rules, systems and patterns—however, the future, for the moment, seems very evasive and exerts its unpredictability" (2020).

Let's see what we can find out about these questions by playing with bet-hedging, seeing where it takes us. If an art world institution—I'll use an art school for this thought experiment—were "bet-hedging" like a bacterial population, what would it do?

Maybe "bet-hedging" means preserving the school archive—which documents variants in curriculum, faculty, and administrative structures—is a priority.

Maybe "bet-hedging" means rooting through that archive, asking pointed questions about how the school functioned differently in past geographically isolated times.

Maybe "bet-hedging" means actively creating a cultural "genome" from which the school could be reborn.

Maybe "bet-hedging" means figuring out what's in an institutional genome . . . and where it would live . . . The Smithsonian? A submarine maintained by a secret society?

Maybe "bet-hedging" means bringing indigenous histories and ideas into play on the site, reaching back to the pre-institutional past.

Maybe "bet-hedging" means feeding all the school's past data to an AI to see what possible structures it generates.

Maybe "bet-hedging" is just a step toward other generative metaphors. In the fluctuating conditions we face, the most urgent reason to stretch for a new metaphor is the stretch, not the particular metaphor. It's part of "adapting" our patterns of thought, opening the possibility to think past "we're going to die" and "this is the way it's done."

Here's the rub: even if we and other artists, schools, galleries, and museums survive the pandemic, the past is still not going to be a good predictor of the future of the global art system—because all global systems will be changing. According to NASA, "Even if we stopped emitting greenhouse gases today, global warming would continue to happen for at least several more decades, if not centuries. That's because it takes a while for the planet . . . to respond, and because carbon dioxide—the predominant heat-trapping gas—lingers in the atmosphere for hundreds of years" (Earth Sciences Communication Team n.d.).

Can artists and art world makers actively practice (keep making a global art system) in the midst of environmental and social turbulence? There is no academic answer to this question; we have to make the answer with our lives. There are cultures that have been navigating catastrophe for a long time, cultures that belong to some of us and from which others of us could, very humbly, learn. But are there also potentials arising from within the existing art system, as it bubbles with change?

Here I turn to the artist and theorist Hito Steyerl. Even before the pandemic, speaking with art critic Despina Zefkili in 2019, Steyerl described the global art system as "losing a sense of international perspective." The system, an assemblage of trade, travel, and discourse that made the art that happened in Lagos or Havana of interest to art worlds in Dubai and Glasgow, and vice versa, was under stress. Steyerl saw this loss as "a wider sign of rampant isolationism," telling Zefkili: "The idea of the global was already massively undermined in the era of neoliberal globalisation, when it became some kind of Potemkin-style cover-up of criminal privatisation in many parts of the world. It is not a surprise that the art world is now faced with a violent—and to some degree necessary—correction towards the local." Continuing, Steyerl remarked, "There is no centre, just aggressively contracting fortresses of local canons, and less connections in-between."

If the global art system was contracting in November 2019, by November 2020 it was dangerously withdrawn. An *Artforum* feature tracking major events that had been canceled or rescheduled racked up eighty-five entries by June 15, 2020 (Artforum 2020). International exhibitions such as the Venice or Gwangju Biennales were canceled or postponed, museums that serve visitors from all over the world closed, art students studying internationally returned to their home countries. There were countercurrents as talks, exhibitions, and trade moved online, but these were only eddies within the current of art activity draining away, perhaps pooling in local studios and art sites but no longer flowing globally.

This is change, but is it a problem? Back in 2019, Steyerl found things to embrace about the breakdown of the global art system, saying: "[T]he art world has been relying for too long on cheap airfares and fossil-based tourism and transportation" and "The artist residency economy needs to be reimagined, too, with its paradigm of enforced nomadism" (Zekfili 2019).

Let's stop with Steyerl's statement for a moment and ask ourselves: How does it make me feel? Does the situation, as Steyerl describes it, threaten something I care about? Maybe I was looking forward to applying for an artist residency? Or am I sick of talking on Zoom and I really would like to be part of the mass migration to see the next São Paulo biennale—if there is one—in person? Or perhaps I would just like to enjoy the odd painting or installation without feeling hectored to change my ways? We're not often asked to question a text in this way, as if we are, as individuals, exactly the subjects in question, the ones whose hopes and desires might be stymied or redirected. Yet if those desires are not acknowledged, they may produce disappointment, anger, and resentment, claiming our attention and keeping us focused on the possibilities of the past rather than looking at the possibilities that are appearing in front of us.

This is not to say that there is no appropriate moment for those emotions, but the life of every animal facing immediate threat depends on perceiving and responding accurately to exactly, specifically, what is present, not to imagined preferable conditions and not to the conditions that were there an hour ago or 150 years ago. We ourselves are part of the conditions we need to perceive accurately in order to build a sustainable art world.

There is yet another evolutionary metaphor embedded in my language here—a figuring of humans as animals whose imperative is survival, layered together with the figuring of the art world as an ecosystem. But with my imperative to give you a chapter rather than a book, I will say only that, as with bet-hedging, what matters most here are not the exact questions I mined from the text or even the text itself—there are many texts I could have used—but the generative approach, the emphasis on questions, always remembering with Steyerl that "the future poses a 100% risk for human health."

Yet we are not helpless. In her book *Territory•Authority•Rights*, sociologist Saskia Sassen, a theorist of globalization, analyzes four centuries of globalization in terms of complex systems, developing a vocabulary for foundational change. Specifically, she speaks of capabilities, tipping points, and organizing logics. Sassen's version of "a capability" is a collective production of a relational system like, say, the capability to organize a massive international exhibition, bringing artistic, social, economic, and physical, logistical networks together to make a statement. One of her key points is that "capabilities can jump tracks and become part of new, emergent, organizing logic" (2006: 144).

In that compact phrase is a key point. For a novel order to emerge from drastic change, certain capabilities have to be preserved, the system has to flex and shift without breaking completely apart. Chaos—a completely disorganized state—readily emerges from foundational change. It is much, much easier to break systems than it is to build them. But Sassen's account of capabilities, tipping points, and organizing logics argues that our current world is the sum total of many prior instances of capabilities "jumping tracks." She gives as an example the advent of European cities in medieval times, writing "the city emerges as a critical institutional environment for the embedding of novel spatial and temporal framings," a change that was enabled in significant part by the fundamental capability that the law mattered (2006: 61).

Something like Sassen's idea of a change that preserves capabilities may be expressed using evolutionary metaphors. If she were using biological rather than mechanical metaphors to describe emergence from capabilities that have "jumped tracks," she might have written of "evolutionary leaps" or "saltations." But however they are expressed, do these analyses or models of large-scale change interact in a revealing way with my questions about change in the global art system? Think, for a moment, about what it means for societies to maintain the capabilities that make the art world hum: the capability to transmit information from experienced artists to novice artists and to audiences, the capabilities to ship freight and move humans that support encounters between art, artists, and audiences internationally, the capabilities of the industrial base that provides abundant materials for art as a side effect of making many other things; the capability to preserve art from other times to inform new audiences and enlarge the cultural archive of information. That is a huge bundle of capabilities. The number of points of possible intervention to maintain it seems overwhelming.

I turn to Hito Steyerl again. In response to a question from Zefkili about the efficacy of institutional critique, Steyerl said: "As no one knows what—if anything—is going to produce any effect, people have to test all options with no guarantees. It's the most rational thing to do. Even making mistakes usually means gaining experience . . . In the long run, this will be even more important than whatever the stated short-term goals are now" (Zefkili 2019).

Her words seem broadly relevant today. When the past is not a trustworthy predictor of the future, we need to help each other "test all options with no guarantees," tolerating uncomfortable uncertainty, something that is very difficult for human groups to do. But it may be the option. In December 2020, art critic John Zeiser wrote: "Global warming is no longer a thing to be averted but a thing to which we must adapt . . . This spring, Arundhati Roy reminded us that pandemics have a way of reshaping the world whether people want it or not . . . Roy described the current pandemic as a 'portal.' She said we could walk through it burdened by the past or go through it lightly" (2020).

The burden of the past is partly inside us. We cannot let it go entirely; we need it to hedge our bets. Equally we need to embrace the unfamiliar in our turbulent environment. There is no secure path to a future. Our choice is to keep making our global art system to the best of our ability right now, through a myriad of concurrent, different mutations that give our metaphoric and actual evolution something to work with. This is hard—but as we saw in the pandemic in so many ways, art calls forth the spirit in people that keeps them going through hard times. So we artists, all of us, have a future. If we make it.

References

Algus, M. (2020). "With the Art World Moving Online, Can a More Egalitarian Community of Galleries Evolve." *ARTnews*, May 14. https://www.artnews.com/art-news/market/digital-galleries-coronavirus-mitchell-algus-1202687171/

Allday, E. (2021). "Bay Area orders 'shelter in place,' only essential businesses open in 6 counties." *San Francisco Chronicle*, March 16. https://www.sfchronicle.com/local-politics/article/Bay-Area-must-shelter-in-place-Only-15135014.php

Artforum (2020). "Art World Corona Virus Tracker." June 15. https://www.artforum.com/news/coronavirus-art-world-tracker-canceled-and-rescheduled-events-82427

Beer, G. (2009). *Darwin's Plots: Evolutionary Narrative in Darwin, George Eliot, and 19th-Century Fiction*, 3rd ed., Cambridge University Press.

Boucher, B. (2020). "Amid Turmoil, SFAI Will Resume Offering Classes in the Fall, Leaving Staff with Concerns." *ARTnews*, August 13. https://www.artnews.com/art-news/news/sfai-resumes-offering-classes-staff-concerns-1202696748/

Brooks, D. R. (2011). "The Major Metaphors of Evolution: Visualizing the Extended Synthesis." *Evo Edu Outreach* 4: 446–52. https://doi.org/10.1007/s12052-011-0360-3

Burnham, J. (1969). "Real Time Systems." *Artforum*, September. https://www.artforum.com/print/196907/real-time-systems-36480

Clarke, B., and Rossini, M., eds. (2010). *The Routledge Companion to Literature and Science*, Routledge.

Darwin Correspondence Project (2020). "Survival of the Fittest," Cambridge University Library, University of Cambridge. https://www.darwinproject.ac.uk/commentary/survival-fittest

Earth Sciences Communication Team, Jet Propulsion Laboratory and California Institute of Technology, "Is it too late to prevent climate change?" NASA: Global Climate Change: Vital Signs of the Planet. https://climate.nasa.gov/faq/16/is-it-too-late-to-prevent-climate-change/

Ghoshal, S., and Bhuyan, A. (2020). "How leading Indian artists are making sense of the covid-19 pandemic." *Mint*, August 8. https://www.livemint.com/mint-lounge/features/how-leading-indian-artists-are-making-sense-of-the-covid-19-pandemic-11596783578749.html

Gigliotti, R. A. (2020). "The Impact of COVID-19 on Department Chairs." *Inside Higher Education*, October 27. https://www.insidehighered.com/advice/2020/10/27/study-documents-how-pandemic-has-exacerbated-challenges-beleaguering-department

Khalil, A. (2012). "As Above, So Below." From Dorion Sagan, ed. *Lynn Margulis; The Life and Legacy of a Scientific Rebel*, Chelsea Green Publishing.

Lorentzen, H. F., Benfield, T., Stisen, S., and Rahbek, C. (2020). "COVID-19 is possibly a consequence of the anthropogenic biodiversity crisis and climate changes." *Danish Medical Journal* 67(5): A205025. https://ugeskriftet.dk/dmj/covid-19-possibly-consequence-anthropogenic-biodiversity-crisis-and-climate-changes

Marzen, S. E., and Crutchfield, J. P. (2018). "Optimized bacteria are environmental prediction engines." *Physical Review E* 98. https://journals.aps.org/pre/abstract/10.1103/PhysRevE.98.012408

Salomon-Fernandez, Y. (2020). "What Community Colleges Need Right Now!" *Inside Higher Education*, December 14. https://www.insidehighered.com/blogs/university-venus/what-community-colleges-need-right-now

Saltz, J. (2020). "The Last Days of the Art World. . .and Perhaps the First Days of a New One." *Vulture*, April 2. https://www.vulture.com/2020/04/how-the-coronavirus-will-transform-the-art-world.html

Sassen, S. (2006). *Territory, Authority, Rights: From Medieval to Global Assemblages*. STU-Student ed. Princeton University Press. http://www.jstor.org/stable/j.ctt7sx98

Sawon, M. (2020). "This Is the Toughest Challenge My Business Has Ever Faced. But Here's Why Small Galleries Like Mine Will Come Out Alive." *Artnet News*, April 27. https://news.artnet.com/opinion/magda-sawon-postmasters-op-ed-1845471

Subramaniam, B. (2014). *Ghost Stories for Darwin: The Science of Variation and the Politics of Diversity*. University of Illinois Press.

Zefkili, D. (2019). "Hito Steyerl: How to Build a Sustainable Art World." *Ocula Magazine*, October. https://ocula.com/magazine/conversations/hito-steyerl/

Zeiser, J. (2020). "How Could It Get Worse?" Brooklyn Rail, December. https://brooklynrail.org/2020/12/field-notes/How-Could-It-Get-Worse

11

Embracing Viral Uncertainty? It's Complicated

Roberta Buiani

As a relatively privileged artist, curator, and academic living in Toronto, I have been fascinated by viral phenomena for many years, to the extent that I have often wondered whether their complex and unpredictable behavior could be endorsed as a remedy against humanity's fixed epistemologies and search for certainty. In this essay, I draw on my personal observations on the COVID-19 pandemic and on two examples in the arts to reflect on how viral uncertainty, polarization, and economic and social disparities have shaken us and simultaneously may help us formulate new approaches to life and to (re)think what it means (and what it could mean) to be human. COVID-19 taught me that in a world dominated by inequalities and opportunism, to "embrace the viral" can be risky, and may occur only when communities, not individuals, stick together and commit to protect, and be in solidarity with, each other.

The Curious Seed

In June 2017, I visited the 57th Venice Biennale. The Danish Pavilion drew my attention with an installation titled *influenza. theatre of glowing darkness* (Roepstorff 2017; *Venice Biennale 2017 Archive* 2017). The term "influenza" in this context stood for the verb "to influence" and for the infectious condition that cyclically visits us. But the artist Kirstine Roepstorff also had something else in mind.

As part of an installation extending beyond the pavilion, merging with the surrounding Giardini, visitors were invited to a short play in a pitch-dark room lit by lights unpredictably bouncing around through what seemed like a maze of glass. The visuals were purposely ambiguous. Roepstoerff did not appear interested in directing spectators toward any specific, pre-established conceptual end point. All she seemed to want from them was to experience darkness, not as a fearful, indeterminate space, but as a place, a moment where growth is possible.

Built as a dialogue between a seed impatient to grow and burst out of its darkness, a dark river, and a midwife, the play invites participants to reflect on aspects thought to stand in opposition but that are in truth two sides of the same coin. Darkness, typically associated with negativity and the unknown, is juxtaposed with light, symbolizing positivity and knowledge. Roepstorff asks us to question these commonplace notions. When the world goes dark, how will we react? Historically, as fear of darkness has been associated with the unknown, with roads unlit (lack of light) and mysterious ailments (microbes) killing humans without apparent reasons, darkness is often unwelcome. By grouping influenza and the fear of the unknown, and by proposing that instead of being overwhelmed by fear, we start searching for those items in darkness that can produce growth and renewal, Roepstorff invites us to embrace the unknown.

When I came across this installation, I found myself rooting for the seed, its impatience, and its youthful spirit of adventure. It was maybe because I have always had a soft spot for the brave and the risk taker, the rebel and the misfit.

But that was 2017. Is it still possible to root for the seed?

Losing Trust, Gaining Trust

As SARS CoV-2 relentlessly spread across the world, it revealed and exacerbated societal disparities, endangering already fragile relationships of trust. Across nations, many were separated for long periods from friends and family. Forced isolation and stressful conditions juggling work and family at home affected the lives and mental health of countless individuals. Many were still obliged to work in unsafe conditions, crammed in warehouses or at the mercy of abusive employers and angry customers. The abrupt cessation of many activities and the closure of businesses caused intense aggravation of poverty and social inequities. The poor became poorer; the sick, sicker (Brown 2020; Buiani 2020; Allen 2020).

At the same time, guidelines from public health officials and epidemiologists became obligations in safety's name. Restrictions were generally interpreted as necessary for the sake of the vulnerable. However, instead of uniting to protect each other, communities became increasingly polarized, basing behaviors on personal circumstances rather than the common good (Buiani 2020; Bird 2020). Those feeling invincible, those in denial, those who said, "it's the elderly who become sick" rejected the rules, as other law-abiding citizens judged them from a distance, calling them names like "covidiots."

As soon as they could, first-world countries resumed their everyday activities, ready to sacrifice their elderly, poor, and vulnerable, but not willing to sacrifice their economies. The rest of the world still struggles to gain access to healthcare and vaccines today. We now see outbreaks of other viral perils, such as new cases of H1N1 influenza and the Marburg virus, as well as threats of future climate change-induced infectious diseases (Nolen 2023; Baker et al. 2022; Lustgarten 2020). Will we ever go back to our now idealized, relatively carefree, and infection-free past lives? Or should we learn to coexist and cohabit with such potentially harmful, but ecologically unavoidable threats (Gorbalenya et al. 2020)?

In a society based on brutal competition, with businesses shuttered, the pandemic left individuals unable to trust each other, prompting them into survival mode. Triage—where medical teams are forced to choose who receives treatment first, "who lives or dies"—is not only practiced within the health sector, but throughout society as a whole (Gonzàlez-Valerio and Martinez Ruiz 2020; Buiani 2020). In this triage mode of thinking, those living in already vulnerable conditions, the immunocompromised, the poor, people with elderly or ill family members, people victimized by racism, and precarious workers were already in danger before the pandemic. Thus, although the pandemic witnessed a surge of generosity and community organizing to help those in need, mutual trust has become rare. As the pandemic has made the world of many individuals more dangerous and insecure, it is their right to seek safety, to judge and call out what they think is unsafe behavior.

Returning to Roepstorff's seed, one side of me maintains a curious and risk-taking attitude toward viral phenomena; the other side feels intense discomfort in adopting such behavior. In a world where privilege, selfishness, and a general disrespect for others make a difference in the outcomes of the pandemic, who can afford to take such risks? How and when can strategies of coexistence with viral phenomena become viable and liberating to the most as opposed to the few? In this current age of viral uncertainty and polarization, how can alternative venues such as arts, community medicine, or citizen science help us better understand medical threats and create community?

Learning from Viral Phenomena: Uncertainty over Confidence

As a first-world academic, viral phenomena have always seemed distant, both temporally and geographically. The flu visits us every year, disrupting lives, but I had personally never examined its impact on different groups of people. However, I have been fascinated by the relative fluidity and unpredictability of the viral, and the similarities with its various concepts.

Due to a series of historical and cultural conjunctures, like the simultaneous rise of the AIDS pandemic and computer network insecurity (Ross 1991), or the reconsideration of molecular biology as an information and communication system since the 1950s (Kay 2000), we now use the term "viral" to define phenomena spreading widely and exponentially, through patterns difficult to predict, exploiting complicated intersections with the surrounding environment or networks (Subramani and Rajagopalan 2003).

The way the viral operates is counter to today's tendency to manage life with increased precision and efficiency. It is incompatible with mainstream paradigms, where health is rationalized through a series of quantified parameters, statistical data, and models to mechanize and simplify everyday activities. However, these models cannot reconcile all variables that converge into complex, long-term, and geographically extensive phenomena. While science acknowledges the flaws of mechanistic models, constantly adjusting and adding new variables, viral phenomena have so many variables and possible outcomes that it seems impossible to predict their outcomes.

I believed that one way to cope with the unruliness and sense of uncertainty that viruses generate was to learn from and about their many facets and the many reactions to them, rather than just focusing on how to fight them with sophisticated technologies. For instance, knowing that the virus is better transmitted indoors could encourage individuals to invent new ways to interact outdoors. However, it takes more than individual initiatives to venture into an uncertain future. As different personal circumstances produce different reactions to safety, health, and trust, a community of care that one can count on for support and collective action must facilitate such a process. Caring, Maria Puig de la Bellacasa explains, is "everything that we do to maintain, continue and repair our world, so that we can live in it as well as possible" (de la Bellacasa 2017). Caring is a collective endeavor: "The question of how we learn to live with others, being in the world, is an opening to 'becoming with'—to be touched as much as to actively touch" (de la Bellacasa 2009). Coping with and embracing the viral must come from a collective act of learning and transforming together. It cannot be performed by the individual alone.

Embracing the Viral and Care

In 2012, Italian artist and computer expert Salvatore Iaconesi was diagnosed with brain cancer. As doctors made suggestions based on standardized treatments, he found himself caught in a medical loop intent to measure, visualize, and examine his condition only. He no longer felt like a human being but "his own tumor" (Iaconesi 2013). His experience with the medical system was a ritual, during which both his body and humanity became just "blood exams, images of body parts, lab values, diagnoses" (Delfanti and Iaconesi 2016).

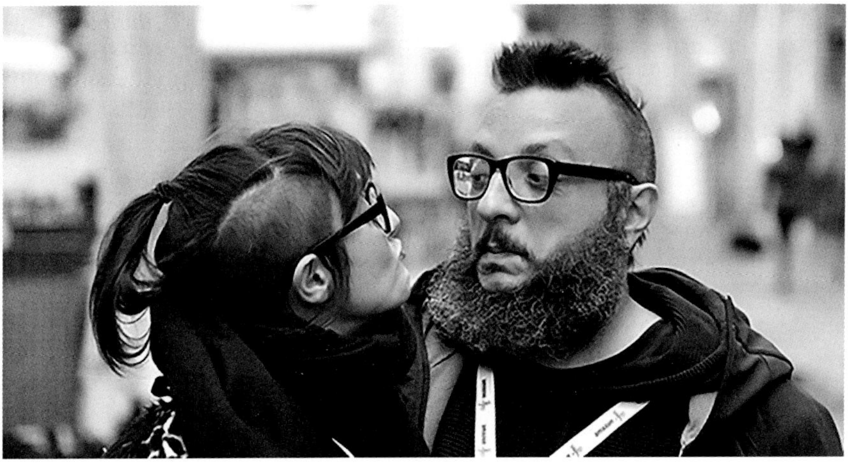

FIGURE 11.1 *Salvatore Iaconesi and Oriana Persico,* HER: She Loves Data, 2020. *Photograph. Courtesy of Oriana Persico.*

Iaconesi decided to break free from the constraints and impersonal approach imposed by the medical establishment through his project *The Cure*, by sharing his medical data online and turning to the community for advice, and comfort. While this action could be identified as a gesture made in the spirit of citizen science, it has a much broader added value, drawn by a need to open up "cancer's source code" as a biopolitical rite of (collective) healing, aimed at redefining concepts, such as "disease" and "cure," "… to re-appropriate the condition of being ill, and to foster a society that recognizes disease as a complex experience—one felt by social bodies as much as individual bodies" (124). Therefore, the gesture is not about hacking or sharing science, but ethics and building a community of (and that) care(s). Iaconesi chose not to rely completely on the anonymous machine that claimed to save his life through scanning, measuring, and probing the body, but on a community he could trust, participate in, and share and feel with. The point was not to save his own life, or to share data to receive more data, but to create a new community of care.

Iaconesi's situation reminds me of the condition of the vulnerable during the pandemic, caught in a society encouraging competition, not solidarity; endless accumulation, not sharing; individualism, not community. Just like Iaconesi's body, seen as data or disease, these individuals are numbers and casualties. But like the impatient seed in Roepstorff's piece opened to darkness and the unknown, Iaconesi let the outside world in. The act of embracing the viral therefore may not be inappropriate after all if this process is achieved as a collective of caring beings. A critique of the Biennale installation reads: "If flu is spread through social contact, its antidote may also be found in its own logic of person-to-person transmission: each individual's ability to make affective choices, the grassroots power to influence change. Influenza is conceived as both symptom and cure" (Roepstorff 2017).

References

Allen, K. (2020). "Lockdown Worked for the Rich, but Not for the Poor. The Untold Story of How COVID-19 Spread across Toronto, in 7 Graphics."

Baker, R. E., et al. (2022). "Infectious Disease in an Era of Global Change." *Nature Reviews Microbiology* 20(4): 193–205. doi:10.1038/s41579-021-00639-z

de la Bellacasa, M. (2009). "Touching Technologies, Touching Visions. The Reclaiming of Sensorial Experience and the Politics of Speculative Thinking." *Subjectivity* 28(1): 297–315.

de la Bellacasa, M. (2017). *Matters of Care: Speculative Ethics in More than Human Worlds*. University of Minnesota Press.

Bird, G. (2020). "Biomedical Apparatuses or Conviviality?" *Topia: Canadian Journal of Cultural Studies*, no. COVID19 Essays (Spring).

Brown, K. (2020). "The Pandemic Is Not a Natural Disaster." *The New Yorker*. https://www.newyorker.com/culture/annals-of-inquiry/the-pandemic-is-not-a-natural-disaster.

Buiani, R. (2020). "Triage Culture: On Losing and (Re)Gaining Trust at the Time of COVID-19." *Topia: Canadian Journal of Cultural Studies* 41: 177–85.

Cepelewicz, J. (2021). "The Hard Lessons of Modeling the Coronavirus Pandemic." *Quanta Magazine*, January 28. https://www.quantamagazine.org/the-hard-lessons-of-modeling-the-coronavirus-pandemic-20210128/

Delfanti, A., and S. Iaconesi. (2016). "Open Source Cancer. Brain Scans and the Rituality of Biodigital Data Sharing." In Barney, D., Coleman, G., Ross, C., Sterne, J., and Tembek, T. (eds.), *The Participatory Condition in the Digital Age*. University of Minnesota Press, 123–44.

Gonzàlez-Valerio, M. A., and Martinez Ruiz, R. (2020). "Triage: Deciding the Ethically Undecidable." *Critical Times*. https://ctjournal.org/2020/05/05/triage-deciding-the-ethically-undecidable/

Gorbalenya, A. E., Baker, S. C., et al. (2020). "The Species Severe Acute Respiratory Syndrome-Related Coronavirus: Classifying 2019-NCoV and Naming It SARS-CoV-2." *Nature Microbiology* 5(4): 536–44. doi:10.1038/s41564-020-0695-z

Iaconesi, S. (2013). "La Cura, an Open Source Cure for Cancer." *Big Data*, August. doi:10.1089/big.2013.0028

Kay, L. E. (2000). *Who Wrote the Book of Life? A History of the Genetic Code*. Stanford University Press.

Lustgarten, A. (2020). "How Climate Change Is Contributing to Skyrocketing Rates of Infectious Disease." *ProPublica*, May 7. https://www.propublica.org/article/climate-infectious-diseases

Nolen, S. (2023). "New Marburg Outbreaks in Africa Raise Alarm About the Deadly Virus's Spread." *The New York Times*, April 3. https://www.nytimes.com/2023/04/03/health/marburg-outbreak-vaccine-treatment.html

OurWindsor.Ca. August 2. https://www.ourwindsor.ca/news-story/10129678-lockdown-worked-for-the-rich-but-not-for-the-poor-the-untold-story-of-how-covid-19-spread-across-toronto-in-7-graphics/

Roepstorff, K. (2017). *influenza. theatre of glowing darkness*. Venezia: Biennale di Venezia. 57. Esposizione Internazionale d'Arte.

Ross, A. (1991). "Hacking Away at the Counterculture." In *Strange Weather. Culture, Science and Technology in the Age of Limits*. Verso, 75–100.

Subramani, M. R., and Rajagopalan, B. (2003). "Knowledge Sharing and Influence in Online Networks via Viral Marketing." *Communications of the ACM*.

Venice Biennale 2017 Archive. (2017). Influenza. Theatre of Glowing Darkness. http://contemporaryartgroup.info/view/venice2017/index.php?id=15274

12

The Anosmatic Symposium

Regine Rapp and Christian de Lutz

Anosmia—the condition of having no sense of smell or being unable to smell certain things.

CAMBRIDGE DICTIONARY, N.D.

Temporary loss of smell, or anosmia, is the main neurological symptom and one of the earliest and most commonly reported indicators of COVID-19.

JIANG, 2020

Introduction

Art Laboratory Berlin (ALB) is a multiple award-winning art and research platform and gallery space, directed by Regine Rapp and Christian De Lutz. Our main goal is the presentation and mediation of contemporary art at the interface of art, science, and technology, often with a biological focus. In this chapter we mark the seasons of the COVID-19 pandemic through the frame of social exchange, both human and microbiological. Originally written in 2021, we offer a curatorial perspective on presenting and discussing biological art both prior to and during the pandemic, including artist Ken Rinaldo's prescient work *Borderless Bacteria / Colonialist Cash*. We share how certain aspects of the pandemic opened up possibilities for exchanges with artists, scientists, and academics across the United States, Europe, and East Asia. In doing so, through the concept of the anosmatic symposium, we use a sensory lens to reflect upon the conflict between the growing ethical imperative to reduce global travel and the basic human need to gather and engage in social exchange.

How have the last few years changed the way we think about exchange? By exchange we mean the exchange of ideas, but also the modes of exchange. For ideas have never been solely fancies of spirit, but also come from physical and metabolic sharing between people. It is no coincidence that Plato's *Symposium* (383–370 BCE) took place during a banquet. The sharing of food and liquids fermented by yeast has always served as a starting point for thought. Before the COVID-19 pandemic, politicians, academics, and artists, many decrying the ravages of global heating, nevertheless spent countless hours flying to attend events, to break bread and drink wine with colleagues as part of global symposia. Most should have and did know better. In this short chapter we ponder, based on our own experiences, how human "habits of exchange" have been altered as a result of the "social interactions" of microbes, and what lessons we can take with us as the pandemic (slowly) subsides.

Borderless Bacteria

On March 1, 2020, we sat in the ALB exhibition space with microbiologist Regine Hengge from Humboldt University, Berlin. It was the last day of the exhibition *Borderless Bacteria / Colonialist Cash*, featuring works by artist Ken Rinaldo. That afternoon, we knew we would be closing our exhibition space for some months, and that our view of the world would radically change before it reopened. The exhibition itself could not have been better timed. Not only was it closing just three weeks before the whole country would shut down, but the exhibition's material and theme, as well as the genesis of the ALB version were eerily on target. *Borderless Bacteria / Colonialist Cash* (Rinaldo 2020) is an exhibition of large Petri dishes, prepared with agar and LB medium, on which two cash bills per dish are pressed. Over the time of the exhibition, the "microbiome" of the money sprouts and grows over the bills and the dishes. Part of Rinaldo's intent was to contrast the open borders of our global economy to the closure of borders for many of the world's inhabitants. While first-world business people, tourists, and artists could travel with few hurdles, this option is limited for many in the developing world, but not necessarily, the artwork shows, for the microbial species they share their bodies with.

Rinaldo's work, as well as his ethical stance, highlight the ways in which gatherings have become problematic on a global stage. In 2019, Rinaldo decided that he did not want to travel to Europe to produce this work. His choice was in response to global heating (referencing Greta Thunberg). For the exhibition at ALB, guided by citizen science approaches that we have used in the past, many of the Petri dishes were produced along with a local secondary school science club. Rinaldo suggested possible combinations of bills from different currencies, but encouraged us to think up our own. We decided to "crowdsource" the money, asking artists, colleagues, and friends to "donate foreign currency bills of low monetary value" to the project. Bills

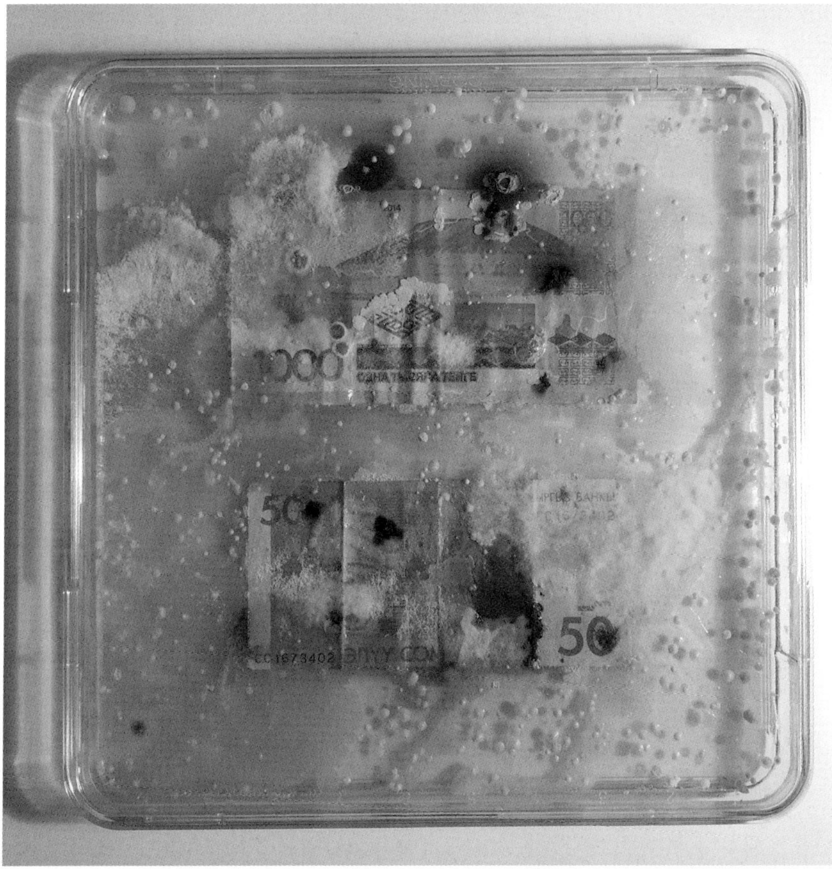

FIGURE 12.1 *Ken Rinaldo,* Borderless Bacteria / Colonialist Cash. *Installation detail. Photo: Art Laboratory Berlin, January 2020. Courtesy of Art Laboratory Berlin.*

came from Europe, the Unites States, China, Vietnam, Brazil, but also Turkey, Ethiopia, Uganda, Kazakhstan, and Kyrgyzstan. Whether the bacteria, fungi, and viruses that grew on the bills came from these countries or nearer to home, we cannot say, but different bills did produce a variety of biofilms. Our age may be increasingly called the Anthropocene, but we still live on a planet dominated by microbes.

Anosmatic Connections

Our afternoon with Hengge was one of our last intimate conversations with a colleague on matters of biological arts and transdisciplinary exchange for several months; at least our last live conversation. And we want to make clear there is a distinction. For a conversation is never solely about words, but always has a thousand nuances, many consciously imperceptible, but nevertheless vital, determined by body language, but also by our body chemistry, the chemicals present in the environment, and of course the effect of sharing food or drinks.

Soon we would be having a multitude of conversations, with colleagues we rarely saw, through a variety of software programs, stylized by webcams and a qualitative variety of internet connections. Yes, sometimes we would lift a glass of wine and toast a friend across the ocean (or across the city) but it was a certifiably different situation. One thing was radically missing—smell! Visually dominant, with sound as an accompaniment, the internet is marked thus far by a neglect of most of all the so called "lower senses." Having meals with friends via Zoom emphasized our very separateness, despite being able to see them in their kitchens and having wonderful conversations. These were anosmatic meals.

That spring, we also wondered what was happening to our microbiomes. Were our holobiontic ecosystems becoming island ecologies (Finlay et al. 2021)? Would this affect our immune systems, once we started socializing again? Autumn 2020 and Winter 2020–1 had extremely light influenza rates. It seems the measures taken to limit COVID—mask wearing, hand washing, social distancing—could not prevent second and third waves of the coronavirus, but led to one of the lightest flu seasons in memory (Peek 2021). We also became aware of perhaps the strangest symptom of infection by the SarsCov2 virus—anosmia—the (usually temporary) loss of smell. We also became aware that anosmia filled our social interactions, which were almost solely limited to online contact (Butowt and von Bartheld 2021; Jiang 2020).

In Spring 2020, we participated in the Viral Culture online sessions organized by Claire Nettleton, which encompassed the nine or so time zones east of California. In part, spurred on by this, we arranged our own *Viral Cloud*, reaching artists across Europe and East Asia, together with Georg Tremmel from BioClub Tokyo. The *Viral Cloud* project (Art Laboratory Berlin 2020c) began as a means of combining both the local and the global. While artists, curators, and scientists contributed short presentations, an important part of the meetings was direct conversation, where we exchanged news from our localities about the pandemic. In Europe there was much talk about masks, and the way it was thought that masks stripped away personality and identity. But participants in Japan, where mask wearing during the winter flu season has been common for decades, argued personality can be seen "in the eyes."

During Summer 2020, in Europe, a belief that the worst was behind us led to an inevitably harsh second wave. In Germany, 20,000 deaths from COVID were recorded from the start of the pandemic through the second week of December 2020; by the end of February 2021 the toll was over 70,000 (Robert Koch Institut 2020; 2021). Many of the culture wars over lockdown, social distancing, and masking that were present in the Unites

States also became part of German political culture. We feared that the pandemic would lead to a "biophobia" that would limit the ability of artists and institutions in Europe to display living microbial artworks. Luckily, we can say, this has not come to pass. On the contrary, knowledge of microbiology has increased if anything, leading to more interest, and less fear involving artworks featuring microbial life.

We had planned a large exhibition on life and care for Spring 2020. Hoping for the best, we postponed, and were able to run *The Camille Diaries. New Artistic Positions on M/otherhood, Life and Care* in Autumn (Art Laboratory Berlin 2020a), closing again just a few weeks before the second lockdown. The precarious situation in Summer 2020 provided us with a number of challenges, but also precious opportunities. We had planned to bring Canadian artist Nicole Clouston to Berlin to prepare, along with Berlin-based biologist India Mansour, her work *Mud (Berlin)*. In twelve custom-made containers, the work presents a unique portrait of Berlin using Winogradsky columns of mud and water from local lakes, rivers, and wetlands. Instead, following the remote-working example set with Ken Rinaldo, ALB staff under Clouston's direction collected mud and water, and filled the columns which bloomed with microbial life and color over the course of the exhibition. Slovenian artist Špela Petrič's installation *Phytoteratology* became a fitting example of the exhibition's overarching theme of care—specifically the way in which artists, scientists, and citizens become experimental loci of care through bioart projects. We have several biologist collaborators in Berlin universities, yet access to laboratories was now severely restricted. Špela has a PhD in Biochemistry, and her work—in which thale cress embryos grow in a bath of chemicals and hormones from the artist's own body—required sterile conditions. Even though we were generously supported by a local DIY laboratory, limitations at their facilities, as Špela had foretold, led to contamination. Throughout the exhibition we had to decide, along with the artist, which containers to remove, or let continue, in varied states of decay.

We had planned a symposium for the opening weekend, but the pandemic made this impossible. We moved to an online format, which we found to be an improvement over our original plans (Art Laboratory Berlin 2020b). Under "normal circumstances" we would have expected a local audience of 50–100, with most of the additional speakers being based in Berlin. Instead we were able to include all exhibiting artists, speakers from the Netherlands, Spain, the Unites States, and a keynote address by scholar and author Astrida Neimanis (based in Sydney at the time). Over 500 viewers watched the symposium live, with a total of 1,640 viewers as of June 2021. Yet for all its success and wonderful moments of exchange, there were no meetings over coffee, lunch, or evening glasses of wine. It was another (if successful) example of what we have come to call the "anosmatic symposium" where we choose technology's abiotic venue, not because it offers a better space, but because it was the only possibility.

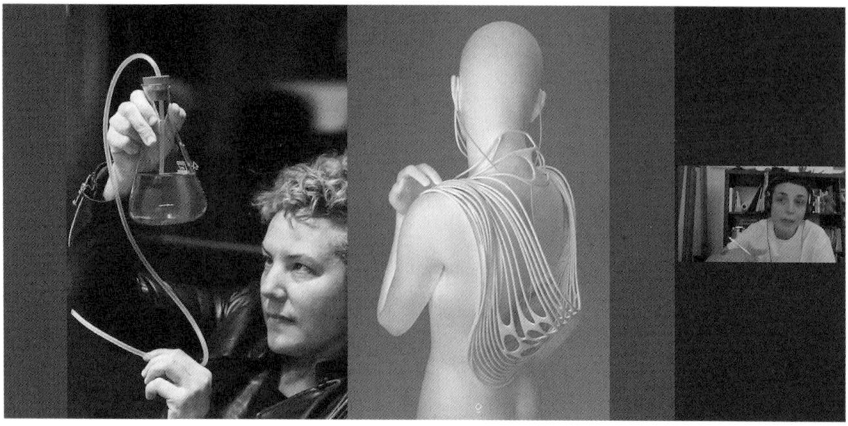

FIGURE 12.2 *Fara Peluso, presenting her talk at VIRAL CLOUD, online forum between Art Laboratory Berlin and BioClub Tokyo, 2020, video still. Courtesy of Art Laboratory Berlin.*

Learning Viral Lessons

For many, the impact of the virus now begins to fade into the background of our daily lives and we once again meet *en masse*, embrace old friends, perhaps even do the ritual *bisous*, that offers an exchange of scent: smelling as a form of greeting. We are contemplating our return to community rituals such as the Do-It-With-Others workshop, which has formed such an integral part of biological and hybrid arts, especially among the generation of artists who have emerged over the last fifteen years. It is not a question of can we "do it together," but how we can do it safely, and without losing the intimacy of communal endeavor. We do not want to give up the lessons learned over the last year and more. It is no longer a question of either/or but how we can use technologies like virtual media platforms alongside a live event. Further with "Zoom fatigue" a concern during the pandemic, how can we use these technologies in a psychologically and biopolitically healthy manner.

We return to the decision that Ken Rinaldo took in 2019. Travel was possible, but neither necessary nor desirable. We hope that travel will be used more thoughtfully in the future. Rather than deny the human need to gather, we may think instead of investing time in our travels. Exhibitions will be nested in residencies, perhaps conferences will be nested into several week-long programs of conferences, workshops, teaching, and learning. Lest we forget, this pandemic arose from an unsustainable relationship between human cultures and the environment, and it is unlikely to be the last pandemic of its type. We are only starting to learn how to live with our microbial colleagues.

Our anosmatic symposium leaves open possibilities to learn from the difficulties of this period, and to use that knowledge for positive outcomes. What we fear most, at this moment, is the very human desire to seek "normality" even if a return to the "normal" of 2019 would be but a *fata morgana.*

References

ANOSMIA | English meaning—Cambridge Dictionary (n.d.). https://dictionary.cambridge.org/dictionary/english/anosmia

Art Laboratory Berlin (2020a). *The Camille Diaries—New Artistic Positions on M/otherhood, Life and Care*. Art Laboratory Berlin. https://artlaboratory-berlin.org/exhibitions/the-camille-diaries/

Art Laboratory Berlin (2020b). *THE CAMILLE DIARIES Symposium—Art Laboratory Berlin*. Art Laboratory Berlin. https://artlaboratory-berlin.org/events/the-camille-diaries-symposium/

Art Laboratory Berlin (2020c). *VIRAL CLOUD | Berlin Tokyo—Art Laboratory Berlin*. https://artlaboratory-berlin.org/events/viral-cloud-berlin-tokyo-iwasaki-rillig/

Butowt, R., and von Bartheld, C. S. (2021). "Anosmia in COVID-19: Underlying Mechanisms and Assessment of an Olfactory Route to Brain Infection." *The Neuroscientist* 27(6): 582. https://doi.org/10.1177/1073858420956905

Finlay, B. B., Amato, K. R., Azad, M., Blaser, M. J., et al. (2021). "Erratum: The hygiene hypothesis, the COVID pandemic, and consequences for the human microbiome (Proceedings of the National Academy of Sciences of the United States of America (2021)," 118 (e2010217118) doi: 10.1073/pnas.2010217118). *Proceedings of the National Academy of Sciences of the United States of America*, 118(11). https://doi.org/10.1073/PNAS.2102333118

Jiang, K. (2020). *How COVID-19 Causes Loss of Smell | Harvard Medical School*. Harvard Medical School News and Research. https://hms.harvard.edu/news/how-covid-19-causes-loss-smell

Peek, K. (2021). "Flu Has Disappeared for More Than a Year—Scientific American." *Scientific American*, April 29. https://www.scientificamerican.com/article/flu-has-disappeared-worldwide-during-the-covid-pandemic1/

Plato (1994). *Symposium*. Oxford World's Classics https://doi.org/10.1093/actrade/9780199540198.book.1

Rinaldo, K. (2020). *Borderless Bacteria / Colonialist Cash*. Art Laboratory Berlin. https://artlaboratory-berlin.org/exhibitions/borderless-bacteria-colonialist-cash/#:~:text=Borderless%20Bacteria%20%2F%20Colonialist%20Cash%20explores,share%20bacterial%20and%20fungal%20communities%3F

Robert Koch Institut (2020). *Täglicher Lagebericht des RKI zur Coronavirus-Krankheit-2019*. Robert Koch Institut, December 8. https://www.rki.de/DE/Content/InfAZ/N/Neuartiges_Coronavirus/Situationsberichte/Dez_2020/2020-12-08-de.pdf?__blob=publicationFile

Robert Koch Institut (2021). *Täglicher Lagebericht des RKI zur Coronavirus-Krankheit-2019*. Robert Koch Institut, February 28. https://www.rki.de/DE/Content/InfAZ/N/Neuartiges_Coronavirus/Situationsberichte/Feb_2021/2021-02-28-de.pdf?__blob=publicationFile

13

FEMeeting

Making of an Antibodies Network

Marta de Menezes and Dalila Honorato

FEMeeting: Women in Art, Science and Technology is the brainchild of artist Marta de Menezes and semiotician Dalila Honorato. Both in our early forties and with family origins in Southern Portugal, we met in 2016 and formed an alliance. We recognized an amazing female presence both at the festivals and conferences that we attend, so we felt this to be the ideal moment to reflect on the status quo of women in contemporary art and science. FEMeeting started with one simple question: What if our girlfriends met?

Originating in 2017, the FEMeeting network was driven by the desire to disseminate projects carried out by women (individuals who identify with the female gender) worldwide and, as a result, to contribute to the development of (a) research methodologies in art and science, and (b) cooperation strategies that can increase knowledge sharing and bring communities together. While an aim of the network is to increase the visibility of women artists and scientists, the moving force of FEMeeting is to encourage collaboration and to promote nonhierarchical gathering spaces (physical and virtual) for individual and collective expression. As FEMeeting gathered significant momentum, the COVID-19 pandemic arose. This essay documents the entwined cultural and physiological senses of the viral that have shaped the course of what FEMeeting is today. We outline the development of this metaphorical "antibodies network," which created resilience, healing, and innovative possibilities.[1]

Contagious by Contact

> Marta: I remember a strong, inexplicable feeling of knowing that the work and the interests of these women would connect. If we could bring them to the same place and give them time to chat, lots of unexpected and unimaginable things would come out of this simple act. I remember when I realized that because we had Cultivamos Cultura, an artist in residency in Portugal that promotes the advancement of knowledge in science, technology, and contemporary art (cultivamoscultura.com n.d.), and we could propose this opportunity to a few of the women we encountered.

> Dalila: I do not remember feeling that this would be too difficult. I just remember the excitement and conviction that this was a great plan to run together. How could we resist the chance to entice these "wonder women" in contemporary art, who were also versed in science? How could we not embrace the pertinence of bringing feminist academia into the safety of a closed-door environment? As we opened the call, word spread like a virus. We understood that sometimes "all women" might not be individually perceived as totally inclusive. Unapologetically, how could we hack the narrative?

Our invitation read: "It will be just women, just us, and we do think this is important for us all and for the unforeseen results of the meeting. We are all women of the same general field of research, of many different ages, and our interests cross. We all have our different methods and research goals, but we do believe that as women we also have an underlying drive that both differentiates us from the rest and brings us closer together as a group."

FEMeeting was initially designed as a traveling experience, with public presentations in urban centers, such as Lisbon, whose talks included innovations to combat the effects of COVID-19 with breathing machines made from blood pressure cuffs which animated Ambu bags (Chacin 2022), to harvesting and processing vaginal fluids to produce DIY beauty kits (Burr-Raty 2019). Another FEMeeting site was Berlin, where, in collaboration with Art Laboratory Berlin, artists proposed novel practices of care in women's and nonbinary individuals' health. WhiteFeather Hunter (see Chapter 14 in this volume) explored menstrual fluid as a means of cultivating stem cells to create a serum nutrient for tissue culture (Hunter 2023). In contrast, the majority of presentations were held behind closed doors in smaller locations like São Luis, Portugal, or Taos, New Mexico. The nomadic duality provided an original structure that enabled social activities at each location, which encouraged communication between participants.

Furthermore, FEMeeting provides strong personal support through online communication, fostering the formation of local nodes to support research and artistic creation. Scheduled for March 2020, FEMeeting Paris was planned as a gathering of women in astronautics and astronomy, while

FEMeeting Troy (New York) was organized to address issues unique to the region's ecosystem and community. As we planned to join these groups, COVID-19 changed the world as we knew it.

> Marta: In January 2020, I was already speaking with my husband Luis (an immunologist) about making a supermarket shopping list to prepare for a lockdown. At the same time, I was planning my annual tour in North America, including workshops in Boston and Buffalo and an exhibition in Toronto before FEMeeting in Troy, New York. At the beginning of March 2020, I was preparing to take my flight to Buffalo, and there was much stress already about COVID, but Europe was ahead of the wave, and there was no real understanding of how things would happen. Then, I had a phone call with Luis in Portugal, and he told me: "You need to come home now, or you are going to spend the next six months in Toronto. Cancel all the programmed activities over the next few weeks, and let's hope this doesn't take too long."

It became evident that we had to cancel the FEMeetings in Troy and Paris. The weight of the situation was crushing, and we were overwhelmed with grief. Tears flowed freely as we grappled with the questions: What was the essence of FEMeeting? What was our purpose as a community? We felt the loss of our cherished dream to gather physically. We paused to contemplate other forms of gathering. Major fans of tea parties and mad hatters, what else if not a mad tea party to cheer us when dystopia knocks at the door?

> Dalila: Meanwhile, I was flying away from Europe into the heart of the pandemic: New York. I chose to isolate myself, with my partner across the Atlantic. It was a dramatic moment of accepting the risk of contagious disease, as well as hoping for a stroke of luck, to remain healthy. For most of 2020, I felt the tug of this weird umbilical cord across continents while sharing the womb that we call FEMeeting. With our sisters, we celebrated life in its strangest extremes and the idea that life continues somehow. Spending hours online became essential, a confessionary, a celebration of individual moments that were everybody's felt accomplishments.

Contamination by Association

Starting in March 2020, we facilitated a nonstructured weekly online meeting—FEMeeting Teapot Chat—to provide support and sanity to those versed in the arts, science, and technology. Hosting the Teapot Chat was a way to maintain emotional stability and discuss the virus's impact in the world around us. With our partners in Paris and Troy, FEMeeting launched a call for creativity, openness, and collaboration in the face of viral contamination and confinement. The result was *Acquired Immunity* (2020), a collection of short videos contributed by various interdisciplinary artists. Contributions ranged from Louise Mackenzie's *Footfall* which chronicles the pandemic's progression through the metaphors of fences and fire during the artist's walk, to Elaine Whittaker's *Touch Coronavirus*, where hands squeeze colorful spiky balls, to Amy Youngs's *Get Grounded*, which gave instruction on how to make soil with worms. Though created in isolation, this assemblage created community and visual manifestations of the virus.[2] They were exhibited online on May 29, 2020, at Ars Electronica Festival as part of FEMeeting Garden, an online exhibition space that hosted videos, artworks, writings, and podcasts (FEMeeting Garden 2020).

In 2021, we started a different type of cartography as we received news of lab space openings led by FEMeeting members, which we celebrated by launching Sister Labs, a series of online watch parties in laboratories. A hybrid model, physical and digital FEMeeting communities were invited. The first announced were Ectopia Lab, in Lisbon, directed by de Menezes; NATURE Lab, in Troy, directed by Kathy High (see Chapter 18 in this volume); and INCUBATOR Art Lab Studio, in Windsor, Canada, directed by Willet. Our hope is to be able to create a safe network of laboratories open to receive and collaborate with women developing research in art, science, and technology.

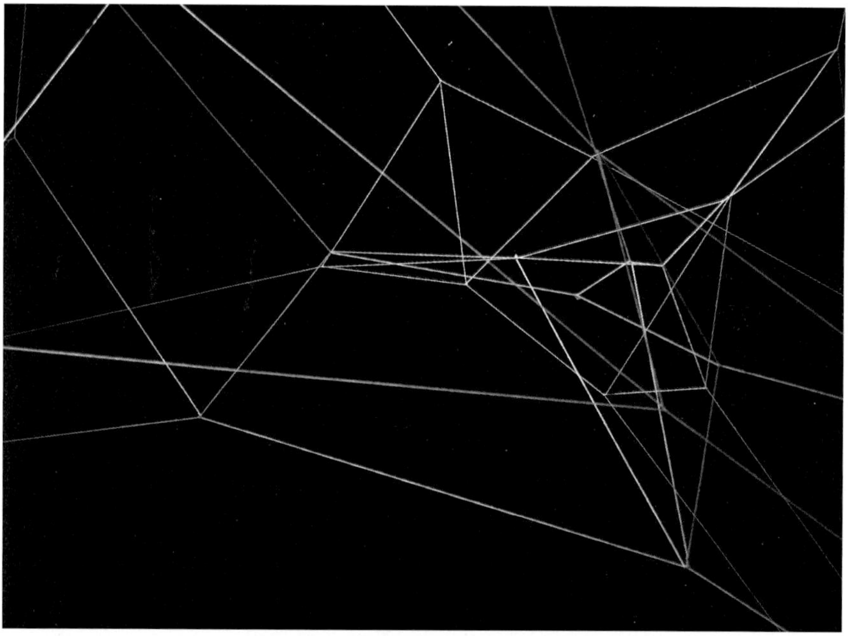

FIGURE 13.1 *Marta de Menezes and Kira O'Reilly,* Spider Web *(2013), installation image. Courtesy of Cultivamos Cultura, Portugal.*

This is how we hack the narrative, by creating nurturing spaces for new forms of being together. Thinking forward and beyond, FEMeeting fosters activities that can have an impact on future generations and that learn from the generosity and wisdom of generations present now. FEMeeting is a complex system of diverse networks that connects its independent members in a web of meaningful connections that will support all the individuals that make up FEMeeting. This web structure is based on the *Web* installation by Marta de Menezes and Kira O'Reilly (2013). FEMeeting thus offers a nurturing and expansive network that allows scholars and artists to thrive beyond traditional constraints.

> Marta: On the first day of our FEMeeting closed sessions, I was struck by the presence of remarkable women, veterans in their respective fields, who generously shared their experiences. Emerging artists and scholars nervously delivered their first public presentations before renowned figures. Emotions overflowed, visible in shaking bodies and quivering voices. A sense of safety pervaded the space, allowing everyone to embrace vulnerability and make mistakes without judgment. In one particular intervention, the presenter broached topics often considered taboo in academic circles with authenticity and courage, prompting a flood of audience support and tears. Through this act of vulnerability, they bestowed upon us an immeasurable gift, reminding us that we thrive despite obstacles, navigating uneven paths as best we can.

> Dalila: With the Teapot Chats, I feel that we were facilitating an exchange of energy with our community, as well as allowing ourselves to feel useful in a time of extreme uncertainty and sense of loss of control over our lives. We drew a circle with invisible chalk and opened the space for whoever wished to join. This was an inclusive gathering, a wider

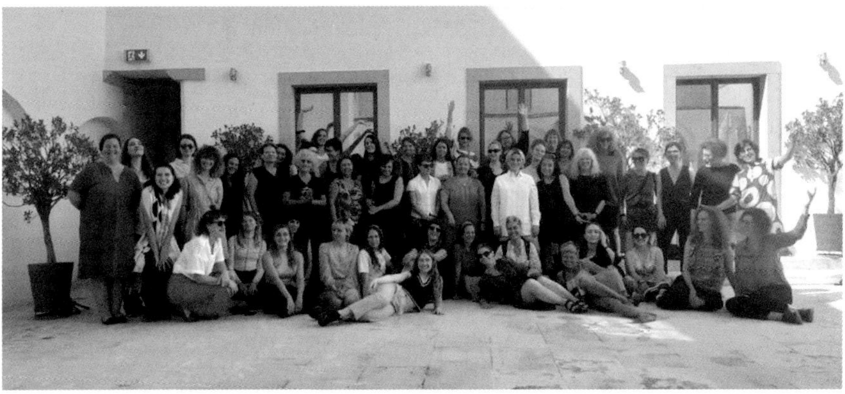

FIGURE 13.2 FEMeeting 2022, Évora, *2022, photograph. Courtesy of Cultivamos Cultura, Portugal.*

continuation of the conversations that Marta and I shared across Europe before 2020. Instead of distressing in isolation we laughed ourselves to sanity and cried ourselves until cleansed.

From sending the trans body to space with Adriana Knouf (FEMeeting 2020: Paris "All Women's Crew" Audio Podcasts) to creating *Acquired Immunity* (2020) videos during lockdown, we have been excited by each other's work and endeavors, and still are, in every FEMeeting. We became interested in each other on a personal and holistic level. Our community created antibody networks during the time of viral devastation, deepening our creative and supportive capacities. It is remarkable to observe that it is no longer just us driving this initiative, which has spread, multiplied, and evolved in like-minded individuals. What else can we ask from a network?

Notes

1 From "Antibodies," the theme of FEMeeting WEB 2020 online event, on May 29, 2020, included: "All Women Crew Podcasts," celebrating women in art, astronautics, and astronomy, and "Acquired Immunity," the video podcasts sent by the members of the FEMeeting community at large in response to the upcoming events brought by the COVID-19 crisis in the first three months of the pandemic. https://femeeting.com/femeeting-2020/

2 All videos available at the Media Sanctuary: https://femeeting.com/acquired-immunity/. Thanks to Kathy High and Brenda Miller.

References

Burr-Raty, I. (2019). *Eco-Errogneous Para Pharmaceutics and Trans-Personal Unisex Cosmetics* in FEMeeting Proceedings, 47. https://femeeting.files.wordpress.com/2021/07/femeeting-2019.pdf

Chacin, A. (2022). *Tethered—The Ventilator Impulse: Eurus, Another Breathing Machine* in FEMeeting Proceedings, 12. https://femeeting.files.wordpress.com/2022/09/femeeeting-book-of-abstract-2022.pdf

"Cultivamos Cultura." *Cultivamoscultura.com*

"FEMeeting 2020: Paris 'All Women's Crew' Audio Podcasts." *Leonardo—Olats*. http://archive.olats.org/space/sasc21/2020/podcasts.php

"FEMeeting 2020: Acquired Immunity' Video Podcasts." *FEMeeting: Women in Art, Science and Technology*. https://femeeting.com/acquired-immunity/

"FEMeeting 2020 Garden" (2020). *Ars Electronica*. https://ars.electronica.art/keplersgardens/en/femeeting/

14

COVID-19 and the Embodiment of Disruption

Assemblages of Agency and the Turducken of Chaos

WhiteFeather Hunter and Molly McKinney

Turducken[1], *n.*
U.S. Cookery.

> A poultry dish consisting of a boned chicken inside a boned duck which is in turn placed inside a (partially) boned turkey, along with seasoned stuffing between the layers of meat and in the central cavity, the whole typically being cooked by roasting . . . (Oxford University Press 2021).

Disease agents intra-act within material bodies and sociopolitical structures to create overlapping patterns of force, reflecting hierarchies in cultural value systems. As they infiltrate situational niches, these agents have the potential to "trick" defenses and amplify themselves. The current result of this amplification is pandemic, causing large-scale system disruptions and breakdowns. Pandemic impacts are measured horizontally in a quantifiable manner, as numbers or percentages of deaths, geographies covered, duration of disease lethality, etc. However, pandemics are most lethal and widespread vertically, greater affecting those positioned lower in socioeconomic/cultural strata. These groups, less able to protect themselves within sociopolitical configurations that value some bodies over others, thus become statistically subsumed. This chapter examines vertical, qualitative impacts of the COVID-19 pandemic by locating nesting agents of disruption in an analogous "Turducken" of Chaos. As a consumable intra-action of bodies—turkey, duck, and chicken—the Turducken serves as meaty metaphor for body politics within a time of pandemic, which we discuss as representative of experiences that traditional medicine hasn't fully considered. Using material feminist analysis with personal recounts, from a patient perspective and the experiences of a frontline medical worker, we illustrate how the system corpus (turkey) invisibilizes female hosts and mediators (ducks) who embody the unruly virus (chicken).

> Like all pandemics, COVID-19 is not an accidental or random event. Epidemics afflict societies through the specific vulnerabilities people have created by their relationships with the environment, other species, and each other. Microbes that ignite pandemics are those whose evolution has adapted them to fill the ecological niches that we have prepared. (Snowden 2019: ix)

Microbes, diseases, and persistent population infections (epidemics) coevolve with humans. Our bodies are not impermeable but fluctuating multispecies relationships—dynamic 'holobionts' (Macfarlane 2019: 14). Medical approaches that bisect the human into a mind–body dualism, separating out "bad" germs from "good" have shifted in recent years; for instance, through now understanding notable impacts of gut microbiota on "individual" mental health (Clapp et al. 2017). Other powerful behaviors of microorganisms, such as spontaneous appearances, rapid spread, and an ability to evolve faster than the pace of research, continue to change the way we perceive these smallest entities. Better-nuanced interpretations of agency among more-than-human actors, within science and the humanities, are transforming our notions of being, survival, and sociality (Crespi 2001). Emergent conversations include expanded notions of bioethics that embrace socioeconomic considerations and intra-acting agencies, forwarded by transdisciplinary researchers within the humanities who work with scientific methodologies (Zurr and Catts 2014; 2018).

In this chapter, we build upon this foundation of transdisciplinary dialogue to present three correlative actors for examination: a novel and wily virus, a "disruptive" human body, and a system of control that carries a violent history of mass disruption. Here, we employ the Baradian term, "intra-act," to pinpoint the dynamism of bodies that amplify their agency through interplay—not simple interactions of inherent agency within already existing actors, but new forms created through forced, complex juxtapositions that might not otherwise occur. As the COVID-19 pandemic unfolds, there have been few times in known history that such a perfect storm of disease agent, carrier body, and systemic conditions have intersected to provide for such rapid paradigmatic change. Specific contexts at the time of COVID-19 outbreak provide a framework for analysis of what we conclude is an opportunity for the debridement of festering dominant societal norms.

Our analogy of the Turducken is salient for two reasons: first, our examination is located in matter and based on a premise that the human body has been reduced to simply meat, "a political anatomy of the body-politic" (McNally 2011: 31). Meat—as worker "units" in the capitalist socioeconomic machine; as research foci for how the novel coronavirus invades body tissues and systems; and as the dead or disenfranchised, unimportant in the eyes of certain publics and governments. The second reason is for the nesting effect of the Turducken, applied toward actors in the pandemic found within one another. There is an immediacy about the insertion of bodies into one another that conveys how these actors behave with and respond to each other.

Our examination focuses on the performance of disruptive "bodies" in the pandemic through a constructed hierarchical lens, from a microscopic-sized virus through individual human female bodies, to societal scale. The virus, as innermost body, is disruptive in our observation of its unpredictable force: causing disease without obvious meaning, and guileful in our negotiations with it. The second body, the human organism, is examined as biologically coded female: disruptive as deviation from hegemonic male standards, confounding medicine with its presentations as "mysterious" and intractable to standardization. The third body is the system corpus of capitalist patriarchy, disruptive in how it perpetuates itself as the controlling force over these other bodies and creates conditions the virus finds useful.

Turducken: The History of Building a Dish

The Turducken, being a large, involved affair constructed of different animals, is gluttonous and excessive, as Western culture is often critiqued to be. Meat as metaphor is potent here because of its overlapping symbolisms. The production of whole animals into segmented meat is a harmful industry that illustrates abjection and devaluation toward bodies not human: "With farmed animals their consumability determines their treatment . . . Attitudes of human exceptionalism, entitlement, and disrespect play a central role in the social rejection of the idea that other animals matter" (Adams 1990: 190). Similarly, human bodies that are not white, male, heterosexual, and/or privileged are also frequently collapsed into the category of "other"—also worth what can be extracted from their component parts.

In preparing our analogous Turducken, to dissect meat bodies that comprise the nested agents, "Turkey'" represents the system corpus, the outermost visible body. This serves as metaphorical embodiment of our cultural paradigm: Western patriarchal capitalism. Capitalism as a system has evolved over time but retains its foundational characteristics: the enclosure and enforcement of compliant bodies to support a labour force that creates wealth for few. In such a system, compliant bodies are material for consumer products, including human products such as organs and tissues for transplant. But also, as we shall show, they serve as repositories of otherness (especially when threatening noncompliance) to be forcefully managed. Forms of violence against animals or "other" bodies is deemed acceptable, even necessary, for capitalist functioning.

In this chapter, meat comes to reflect specific neglect that healthcare, governments, and fellow citizens show toward those most affected by (and least protected from) the pandemic. In examples we discuss: the labouring body that fills an employed position at the hospital, or the body that catches the illness and works to recover at home, it becomes clear that some bodies are considered more human/important than others and are better protected by our social systems. The pandemic has illustrated sweeping systemic disinterest in saving those already least protected, who are treated as "terminal animals" (extraction end points) expected to continue performing self-sacrificial functions for the good of the economy and society (Adams 1990: 190).

Our interest is in the legacy of institutionalized repression of the physiologically female body, focusing on current iterations. Situated in the middle position of "Duck," we explore how the perceived female serves as locus of control for perpetuation of patriarchal capitalism during pandemic contagion—via medicalized disenfranchisement. Patriarchal capitalism has always been part of scientific medicine. Government legislation, corporations, and academia exhibit biases through categorizations of race, class, gender, and sexuality toward affinities for profitability/predictability held by white heterosexual males (Bartky 2003; Roberts 2011). The system corpus in which our deviant bodies are found is reinforced through interchange of such biases.

Patriarchal capitalism also relies upon nonsentient bodies and entities—those named and studied as behaving in a fixed or predictable manner. According to convention, these bodies yield their secrets and weaknesses to the rectifying tools of Western science. Its institutions are invested in the programmed performance of each body located within it, to perpetuate itself in solutionist ways that benefit those holding power. Viruses—here represented by "Chicken"—thought to be unintelligent, are nonetheless slippery, variegated beings coevolving with the human evolutionary trajectory (Welsh and Allman Updyke 2020). Many viruses are resistant to Western medicine's orderly franchise, capable of creating far-reaching effects in the corpus(es) they inhabit.

The bodies we discuss are nested (but leaking) within specific discourses and contexts: the virus within the human, the human within the system, and the system enclosing it all. They intra-act and react to one another, functioning outside of or in resistance to each other. "Deviant" bodies, which we examine through individual case studies, are used to hold up

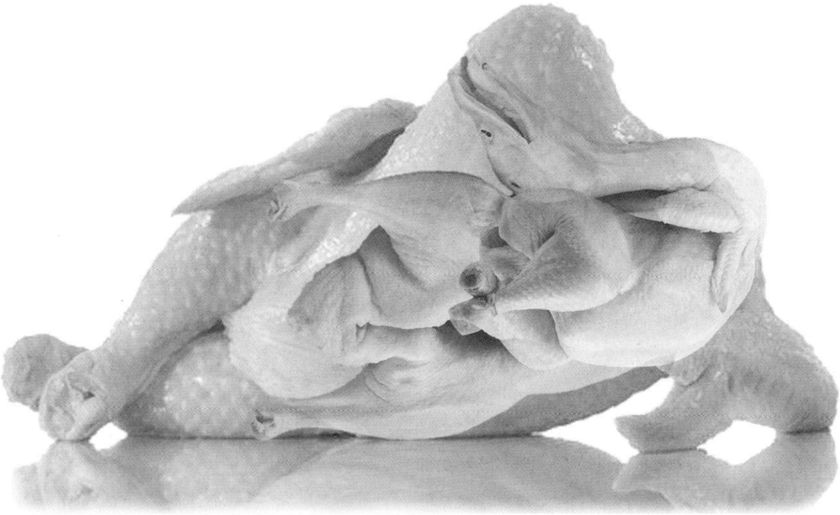

FIGURE 14.1 *WhiteFeather Hunter, 2021,* Digital montage of found photos of raw turkey, duck, and chicken bodies. *Courtesy of WhiteFeather Hunter.* © *WhiteFeather Hunter.*

systems in which they are located. "Labeling [deviance] is largely a matter of some persons or groups imposing their rules on others" (Schur 1984: 6). The deviance of *others* upholds the normality and powerful cultural location of *some* (Weinstock 2020: 1–56). However, as we conclude, deviants can pluck opportunities for transformation from impending collapse.

Accidental Agents, the Ushers of Chaos

Radical changes occur through the body; this site of possible oppression affords insight into one of our greatest resistance abilities. Material feminisms have located the body as an intelligent and autonomous subject capable of resistance. This agency extends to nonhuman bodies via "theories in which nature is more than a passive social construction but is, rather, an agentic force that interacts with and changes the other elements in the mix, including the human" (Alaimo and Hekman 2008: 7).

Much of the change incited by the pandemic seems incidental: the virus causing COVID-19 has not "intended" to act as radical social change agent. Rather, the niche humans created for it allowed it to exploit our weaknesses, as a society and as ill bodies. The virus defied its anticipated performance as a tractable disease within known scientific paradigms: first, it became apparent that there was no reliably consistent pattern to predict the disease trajectory in the human body. Because of how the virus replicates, utilizing cells that follow the angiotensin-converting enzyme 2 (ACE2) pathway and as a result touching nearly every system in the body, illness can manifest anywhere (Sriram, Insel, and Loomba 2020). Also, some illnesses made manifest by the virus followed patterns of "occult" (as labeled in medical terminology) disease: those which remain hidden prior to infection (Paolucci et al. 2021).

Similarly, societal stopgaps designed to respond to outbreaks of pandemic proportions had fissures that remained conveniently out of sight for those privileged enough to ignore them. The virus not only caused disruption to our wellbeing, but also to healthcare and other social systems which provide care. Just as female-coded bodies already disrupt the patriarchal system by deviating from androcentric norms, female-coded job assignments (such as nurse) disrupted the efficiency and productivity of the capitalist system: frontline workers demanding personal safety and wellbeing in the workplace are labeled as radicalized when they threaten to walk out from their jobs. Their demands translate to a demand for global equity in caretaking profession(s). With systemic collapse a possibility in the face of these deepening pressure cracks, the virus acted in similar ways in the human body to the destructive pathways of the system corpus and called into question the supposed objective rationality of the medical system.

Medical science was not always as abstracted from magical thinking as it seems. Diagnoses once revolved around the supernatural: "As a variant, a person could be deemed to be innocent but temporarily 'possessed' by an evil spirit" (Snowden 2020: 14). This illustrates concepts of disease as unnatural and external to society. It posits innocence (lack of knowledge), purity (lack of experience), or other passive forms of "sick" presentation as grounds for pardon. Those who performed health in the approved manner were forgiven when they became sick, not at fault themselves but possessed by a transient affliction. Those who performed health incorrectly were, and still are, considered always-sick (irredeemable) and used as life lessons to others, to learn how to avoid lifetime suffering (Starr 1982: 21). In our case studies below, we discuss how such a precondition for restoration and maintenance of wellness overshadows who receives medical attention during our present health crisis, and who is left flailing.

We currently see this hierarchy and coding of bodies play out in real time, in those deemed worth saving, and those not so much. For example, the COVID-19 Data Tracker (2020) shows that the greatest rate of death from this pandemic is seen in nonwhite populations. Research on which bodies are most susceptible to infection, frequently dying or most likely to experience the long-term disabling effects of the virus, points again and again to oppressed populations as bearing most of the disease burden. Women, as an assigned gender group, and nonwhite women most of all, bear the greatest burden across all groups (Madgavkar et al., 2020).[2] These bodies experience societal contexts that place them at greater risk, such as holding "essential" jobs or being unable to quarantine due to risk of lost wages (The Lancet

2021). From an experimental standpoint, these can be seen as test subjects: female bodies matter for the information their illness generates for others. As ill-performing figures, they prop up the operational status of the system corpus, which in turn refuses to alleviate their suffering. The culturally baked-in "deviance" of the female body provides an excuse for automatic dismissal of their experiential knowledge of disease. Likewise, caretaking individuals are perceived as dispensable within patriarchal capitalism, and their wellbeing easily neglected. The governing system relies on popular narratives that healthcare jobs are inherently dangerous and that caretakers signed up for this danger, like foot soldiers. Meanwhile, it allocates fewer resources to areas of nonpriority (such as provision of adequate personal protective equipment). A survey conducted by the American Nurses Association found that 43 percent of nurses made their own personal protective equipment (PPE), and 14 percent made their own supplementary PPE, such as face masks and hand sanitizer (American Nurses Association 2021a).

The patriarchal capitalist system of medicine, commerce, and social reproduction originally disrupted and continues to disrupt ecosystems and agents that preexisted it. Its attempts to control bodies within certain limitations toward profitable outputs created the perfect storm for uncontrollable disease to arise. Deviance, where predictability is expected, such as with the behavior of the virus causing COVID-19, lodges grit into the cogs of our social machinery. This tiny speck, which brought everything (from local workplaces to global trade routes) to a grinding halt, illustrates the probable breakdown of an inadequate, morally detached, inflexible system. Such a system, reliant on fixity and met by a well-suited biological agent, fostered the system collapse we are seeing now (Snowden 2020).

Disruptions in Seeking or Providing Care: "Catastrophizers" and "Hysterics"

Care, at its foundation, is situational. It "implicates different relationalities, issues, and practices in different settings" (Puig de la Bellacasa 2017: 3). For the purposes of our arguments, care (or the lack thereof) is at the crux of sociopolitical control strategies that we aim to unpack, layer by meaty layer. Who gives or receives what kinds of care, and when, under what circumstances, during the pandemic? What gets stuffed out of sight when embodiments of disease disrupt conventional health and social care? What creative repackaging of bodies is presented to the public to impress that "everything is taken care of" (under control)? Or, as the popular Québec pandemic adage soothed, "*Ça va bien aller*," which translates as "It's going to go well," or everything will be fine. This mantra is reflective of the Second World War British saying, "Keep calm and carry on," implying that society continue to function by ignoring the crisis and maintaining norms of labour matrices at all costs. This captures attitudes of "care" around COVID-19 as vehemently adherent to a traditional Western paternalistic (capitalistic) structure. This structure causes its own damage and has done so since its inception. In two following cases, we examine how healthcare structures in place in Western medicine serve not to help, but to extend harm.

Case Study 1: Healthcare Worker as Catastrophizer (Molly McKinney)

Medical practice, hospital systems, emergency management, and modern nursing were built or refined during wars or other historical episodes of destruction, due to infectious disease, physical altercation, or accidents (Corbett Bell 2009; Tortorice 2017; Hampton 2017). Doctors, nurses, and other healthcare workers were valorized as "soldiers" and heroes. When healthcare functions in a militarized context during crises, healthcare workers overall are expected to immolate themselves in the cause of new warfare. When COVID-19 hit, I went from being a nurse to a "hero."

In Western culture, a hero disregards personal bodily safety to protect others. Fictionally, these characters are treated with adoration and awe, and usually have weapons or superpowers to help transcend their fragile humanity. They are individual actors who rarely need help from other people. However, when a nurse deviates from expectation and refuses to define herself as soldier or hero, problems can arise. I have several chronic conditions that greatly impact my daily existence, one of which is bronchitis. When I was exposed to SARS-Cov-2 at work, my insistence upon the gravity of its possible repercussions for my patients and myself, my requests for help, and my interrogation of systems of safety were treated as hyperbole. My expressions were reflected, however, by nurses countrywide and the American Nurses Association released an updated code of ethics for the pandemic that voiced these concerns and provided guidelines (American Nurses Association 2021b). My workplace became more supportive; however, initially I faced disbelief, a doubling-down on exploitative communications of "duty" and the insistence that I had "signed up for this." I was told that I had a professional obligation to take care of patients with any number of contagious conditions, no exceptions. I was told that staffing safety was a priority and I was to stop asking questions. I was told that, as a nurse, I was essential, a hero.

My body as healthcare worker acted as both a potential reservoir for the virus and problem for the system corpus. Bodies are required to combat the virus on the frontlines of the war against the pandemic. The greater cause to which I was called to sacrifice myself was one I questioned; I criticized its operations, problematizing my position as a nurse who promised to take care of others, and my position as a soldier needed in battle. As with any war, dissent was treated as betrayal (Rosenfield and Burnell 2021). Amid my cognitive dissonance, I chose dissent. I decided that I am not this type of hero, and no one should have to be. For "the 'ethics' in an ethics of care cannot be about a realm of normative moral obligations but rather about thick, impure, involvement in a world where the question of how to care needs to be posed" (Puig de la Bellacasa 2017: 6). As a nurse, it is my job to care for others but also for myself, and to strive to improve the systems of care in which I work (Kreitzer and Koithan 2014: 3). Nursing organizations around the country took this stance, asking nurses to speak out about refusals to sacrifice themselves within harmful cultural locations. The #notahero social media campaign by Trusted Health (2021) was a call for better systems of care.

The people we have been calling heroes are those paid the least and treated the worst (Chagla et al. 2021). We, as a society, find them expendable while paradoxically relying on them for safety. As a nurse, I advocate for the vulnerable and oppressed and speak up for people who do not have a voice. I will care for the sick and injured, and for myself. This is my ethics of care. I will not be the unquestioning, self-sacrificing hero demanded of me, a new soldier against the preventable pandemic. I will not throw my body down for imposed warfare in service of exploitation. That kind of care is not ethical. I am a nurse, #notahero.

Case Study 2: Patient as Embodiment of Viral Deviance (WhiteFeather Hunter)

What does it mean to embody COVID-19—as its antagonized flesh host and internally erupting evolutionary co-construct? Its bedraggled survivor and emotional mouthpiece in a global culture of ambivalence? The puzzling embodiment of the disease is the expression of its deviance, particularly with a viral vector not fully determined by science, as with COVID-19. Such an embodiment has strong implications for how it is perceived, managed, and how one is placed into or receives care.

I contracted SARS-Cov-2 in March 2020, the week before global lockdowns began. At this stage, the viral infection was poorly understood within a narrow set of infection parameters: international travel, contact with a confirmed case, and fever. As an atypical COVID-19 patient, presenting with symptoms not recognized by global health authorities at a time when testing was not widely available, I lacked "proof" of the disease. In cases like mine, where the disease timeline also lingered, care was not forthcoming nor sustained.

Atypical patients may be positioned as a challenge to the authority of medical experts for simply insisting that they are, in fact, unwell. I, along with a mass of self-described "long-haulers" were repeatedly reductively mislabelled as simply "anxious" (Chuck 2020). This insinuation, or sometimes direct assertion, infers that the weird spate of seemingly disparate symptoms is psychosomatic or overblown, and therefore medically resolved; such patients are shuffled out of the hands of doctors, as our chronic conditions no longer warrant medical investigation.

As my disease unfolded in advance of medical understanding, doctors denied that my symptoms were associated with the raging pandemic and looked to explain them within already defined pathologies, particularly the catch-all condition of "anxiety." This was due in part to having been denied testing when I contracted the virus, leaving me with only my own testimony deemed worth nothing. I was informed that I should seek a psychologist to treat the unprecedented and sudden onset of nerve tremors, alarmingly low blood pressure (70/50), lack of urinary retention, body temperature dysregulation (down to 35.2°C), tachycardia and palpitations, prolonged chest pain (following pericarditis), superficial blood clots, difficulty breathing, low blood oxygen levels (down to 79 percent), severe headaches, sinusitis, inability to recall familiar words, lack of coordination, dizziness, neuralgia, joint pain, numbness in my extremities, tinnitus, phantosmia, vision disturbances, exhaustion, hair loss, panic attacks, menstrual irregularities and more—physiological symptoms now reported to occur as a variable cluster in COVID-19 long-haulers (McCorkell et al. 2021).

The fact that the majority of COVID long-haulers are women serves to highlight the inequity of this disqualification for care (Seang, Tubiana, and de Truchis 2020). Like classic misogynist hysteria narratives that take precedent from Western medical history, disbelief and diminishment continue in multiple formats today (Scheurich 2000). One cannot adequately assess this history without examining its patriarchal religious underpinnings: aside from the first Western convalescence facilities being church-controlled sites, concepts of suffering within religious beliefs bleed into current perspectives. Mysterious illnesses that suddenly appear and afflict humans, especially women, have been characterized differently under different conditions: those who quietly bear the trials of tortuous maladies have been deemed divine vessels of grace (Beers 1992). Conversely, women insistently vocal with inexplicable (especially chronic) complaints, demanding or begging remedy, are gaslit as hysterical—even demonic—and deserving of suffering they've brought upon themselves.

The location of the uterus as source of all trouble etymologically links its bearers to the capacity for hysteria as defined by early translation: from *hustéra*, Greek for "womb" (Oxford University Press 2021). Though the medical community promotes objective approaches based on modern science, feminist public health research has established that discriminatory sex and gender biases continue to exist in disease treatment (Jackson 2019). Disdain in the voices of (male) doctors I desperately sought care from contributed to the trauma I experienced and compounded my issues, leaving me vulnerable to emerging, life-threatening manifestations of the disease.

During my multiple dealings with perplexed medical professionals, it struck me that they are the ones with anxiety (understandably so, as the disease progressed overwhelmingly). In a survey conducted during the height of the pandemic, 87 percent of nurses reported fear of going to work, related to short staffing, lack of PPE, and lack of training (American Nurses Association 2021b). Due to the hierarchical nature of Western medicine, where final authority rests with the doctor and not the patient experiencing disease, projecting anxiety onto the patient when no other diagnosis is available enables a clean labeling of the problem as *outside* of the system. The expectation of hysteria/anxiety includes expectation that women accept therapies that may be inappropriate (i.e., antidepressants or psychotherapy); unwillingness to do so is additionally labeled as "difficult," resulting in refusal to offer further care. My assertions that I didn't need antidepressants or anxiety medications were manipulated by doctors through interrogation such as, "What are you afraid of?" or insistence that I take the prescription "just in case" I changed my mind (as if I did not really know myself as well as they did). The only thing I have been truly afraid of is suffering a life-threatening pandemic disease while not being taken seriously.

A Chaotic Conclusion: Squirming into Dissolution and Reform

The novel coronavirus performs autonomy differently within each body it encounters, metamorphosing with each instance of disease. Showing no concern for behaving as it ought and resisting efforts at control, it threatens the foundations of our fragile, violently upheld institutions. It performs this much the same way it navigates internal systems of the body to exploit the most vulnerable points. In this manner, enactments of disruption occur at multiple levels: in the perceived occult agency of the virus (Chicken), in the deviantized body of female patient or caregiver (Duck), and in the bloated system corpus that constitutes our enveloping cultural body (Turkey).

We have discussed how the patient body problematizes predictive medical modeling by acting in concert with this intractable, individual performance of viral disease. Current scientific knowledge, based on amalgamated (and race-/gender-biased) research data, fails to adequately describe symptomatology among patient populations outside of the norm. Those who present in a manner not specified by these standards become deviant by default, noncompliant within a larger system that continues to dismiss them. We have likewise discussed how the caretaker's body becomes problematic, such as through nurses' strikes and news stories that expose working conditions on the frontlines. Bodies that fill predominantly female-coded job roles function as locations for exploitation in our sociopolitical system; the push of expendable bodies toward sites of danger have reframed these people as "essential workers" and "heroes," yet they face enforcement of their positions through threat of unemployment, and resultant loss of housing and health benefits. In both examples, dissent forms specific suppurations that erupt around social inequalities.

When faced with disruption at the level caused by COVID-19, patriarchal capitalist structures, tenuously held in place, are threatened with dissolution. The Turkey, deboned, collapses without the performative support of the other bodies, its flaccidity revealed as deceptions of fixity and omnipotence. We propose that it is time for this dissolution—the Turducken is ripe with odoriferous signals. We cannot discuss the materiality and manipulation of intra-acting "meat" forms as nexuses of sociopolitical collapse without including the squirm of reform agents: so, we leave you with the Maggot. This larval creature performs the communal act of consuming the corrupted—and, in our meaty metaphor, can serve as (counter)force of destruction to rotten systems that entrap bodies within the Turducken of Chaos. The under-celebrated Maggot devours putrescent tissue, excreting protein-dissolving enzymes that inhibit further microbial growth (Abdel-Samad, Mohammad 2019; Yan et al. 2018). Thus, Maggot can be intentionally introduced to festering flesh, such as our Turducken has become, to consume it in a wriggling mass of shining bodies (called debridement therapy).

We suggest that it is time for the debridement of patriarchal capitalism, which has too long pinned us into this meat straitjacket. We call the Maggot to assist in acts of decomposition as we allow the destruction of harmful systems, and radically rethink concepts of illness, care, and embodiment. In our collective action of overturning fallen structures that previously upheld the Turducken, we discover Maggot already swarming from the previously hidden anatomies, in a writhing reanimation of flesh. This is our point of transformation: to employ the monstrosity of decay toward dissolution of the no longer useful. Within the moist stink of the spoiled system, we (as Maggot-ridden) must allow the digestion of what is revealed to be rotten, to facilitate a complete breakdown of what was already always corrupt(ed).

Notes

1 WhiteFeather Hunter would like to acknowledge the support of her PhD supervisor, Ionat Zurr, in providing valuable feedback during the formulation of this text.

2 Racialized groups bear the greatest social burdens regarding pandemic mismanagement, as with the rest of the history of Western society. Examining this issue in detail is beyond the scope of this paper and we, as white women, cannot adequately speak to these experiences.

References

Abdel-Samad, M. R. K. (2019). "Antiviral and Virucidal Activities of Lucilia Cuprina Maggots Excretion/secretion (Diptera: Calliphoridae): First Work." *Heliyon* (5)11: e02791. doi:10.1016/j.heliyon.2019.e02791

Adams, C. J. (1990). *The Sexual Politics of Meat: A Feminist-Vegetarian Critical Theory*. Bloomsbury.

Alaimo, S, and Hekman, S. (2008). "Introduction: Emerging Models of Materiality in Feminist Theory." In Alaimo, S., and Hekman, S. (eds.), *Material Feminisms*. Indiana University Press, 7.

American Nurses Association (2021a). "Nurses, Ethics, and the Response to the COVID-19 Pandemic." https://www.nursingworld.org/~495c6c/globalassets/practiceandpolicy/work-environment/health--safety/coronavirus/nurses-ethics-and-the-response-to-the-covid-19-pandemic.pdf

American Nurses Assocation (2021b). "What 32,000 Nurses Want You to Know About Treating COVID-19." https://www.nursingworld.org/~4987e5/globalassets/covid19/ana_covid19infographic_dataset1_20200424-final.pdf

Bartky, S. L. (1988). "Foucault, Femininity, and the Modernization of Patriarchal Power." In Weitz, R. (ed.), *The Politics of Women's Bodies: Sexuality, Appearance, and Behavior*, 2nd ed. Oxford University Press, 25–45.

Beers, W. (1992). *Women and Sacrifice: Male Narcissism and the Psychology of Religion*. Wayne State University Press.

Catts, O., and Zurr, I. (2018). "Artists Working with Life (Sciences) in Contestable Settings." *Interdisciplinary Science Reviews* 43: 40–53. doi: 10.1080/03080188.2018.1418122

Chagla, Z., Huiting, M., Sander, B., Baral, S. D., and Mishra, S. (2021). "Characterizing the Disproportionate Burden of SARS-CoV-2 Variants of Concern Among Essential Workers in the Greater Toronto Area, Canada." *MedRxiv*. https://doi.org/10.1101/2021.03.22.21254127

Chuck, E. (2020). "These Women's Coronavirus Symptoms Never Went Away. Their Doctors' Willingness to Help did." *NBC News Digital*. https://www.nbcnews.com/news/us-news/these-women-s-coronavirus-symptoms-haven-t-gone-away-doctors-n1235091

Clapp, M., et al. (2017). "Gut Microbiota's Effect on Mental Health: The Gut-Brain Axis." *The Journal of Clinical Practice* (7)4: 987. 10.4081/cp.2017.987

Corbett Bell, R. (2009). *The Ambulance: A History*. McFarland & Company, Inc.

Crespi, B. J. (2001). "The Evolution of Social Behavior in Microorganisms." *Trends in Ecology and Evolution* 16(4): 178–83.

Jackson, G. (2019). "The Female Problem: How Male Bias in Medical Trials Ruined Women's Health." *The Guardian*, November 13. https://www.theguardian.com/lifeandstyle/2019/nov/13/the-female-problem-male-bias-in-medical-trials

Kreitzer, M. J., and Koithan, M. (2014). *Integrative Nursing*. Oxford University Press.

The Lancet (2021). "The Plight of Essential Workers during the COVID-19 Pandemic." *The Lancet*. https://www.ncbi.nlm.nih.gov/pmc/articles/PMC7241973/

Macfarlane, R. (2019). *Underland: A Deep Time Journey*. W.W. Norton & Company.

Madgavkar, A., et al. (2020). "COVID-19 and Gender Equality: Countering the Regressive Effects." *McKinsey and Company*. https://www.mckinsey.com/featured-insights/future-of-work/covid-19-and-gender-equality-countering-the-regressive-effects#

McCorkell, L., et al. (2020). "Report: What Does COVID-19 Recovery Actually Look Like? An Analysis of the Prolonged COVID-19 Symptoms Survey by Patient-Led Research Team." Patient Led Research Collaborative.

McCorkell, L., et al. (2021). "Patient-Led Research Collaborative: embedding patients in the Long COVID narrative." *Pain Reports* 6(1): e913–e13.

N.A. (2021). "The COVID Racial Data Tracker." *The Covid Tracking Project at the Atlantic*. https://covidtracking.com/race

N.A. (2021). "hysteric, n. and adj." *OED Online*, Oxford University Press. www.oed.com/view/Entry/90639

N.A. (2021). "#notahero but a. . . campaign", *Trusted Health*. https://www.instagram.com/p/COxwRM0MTgS/?utm_source=braze&utm_medium=email&utm_campaign=Lead+-+The+Handoff+Newsletter+-+05.14.21&utm_content=mktg

N.A. (2021). "turducken, n." *OED Online*, Oxford University Press. www.oed.com/view/Entry/273637

Paolucci, S., et al. (2021). "EBV DNA Increase in COVID-19 Patients with Impaired Lymphocyte Subpopulation Count." *International Journal of Infectious Diseases* 104: 315–19.

Puig de la Bellacasa, M. (2017). *Matters of Care : Speculative Ethics in More Than Human Worlds*. University of Minnesota Press.

Roberts, D. (2011). *Fatal Invention: How Science, Politics, and Big Business Re-Create Race in the Twenty-First Century*. The New Press.

Scheurich, N. (2000). "Hysteria and the Medical Narrative." *Perspectives in Biology and Medicine* 43(4): 461–76. doi:10.1353/pbm.2000.0054

Schur, E. M. (1984). *Labeling Women Deviant: Gender, Stigma, and Social Control*. Random House.

Snowden, F. (2019). *Epidemics and Society: From the Black Death to the Present*. Yale University Press.

Sriram, K., Insel, P., and Loomba, R. (2020). "What is the ACE2 Receptor, How is it Connected to Coronavirus and Why Might it be Key to Treating COVID-19? The Experts Explain." *The Conversation*. https://theconversation.com/what-is-the-ace2-receptor-how-is-it-connected-to-coronavirus-and-why-might-it-be-key-to-treating-covid-19-the-experts-explain-136928

Starr, P. (1982). *The Social Transformation of American Medicine*. Basic Books.

Weinstock, J. A. (ed.) (2020). *The Monster Theory Reader*. University of Minnesota Press.

Welsh, E., and Allman Updyke, E. (2020). "Episode 43: M-m-m-my coronavirus." *This Podcast Will Kill You*. https://thispodcastwillkillyou.com/2020/02/04/episode-43-m-m-m-my-coronaviruses/

Yan, L., et al. (2018). "Pharmacological Properties of the Medical Maggot: A Novel Therapy Overview." *Evidence-Based Complementary and Alternative Medicine*. https://www.hindawi.com/journals/ecam/2018/4934890/

Young, A. (2020). "Episode: 212 Dr. Vandana Shiva on Becoming Untameable." *For the Wild Podcast*. https://forthewild.world/listen/dr-vandana-shiva-on-becoming-untameable-212

Zurr, I., and Catts, O. (2014). "The Unnatural Relations Between Artistic Research and Ethics Committees: An Artist's Perspective." In Macneill, P. (ed.), *Ethics and the Arts*. Springer, 201–10.

15

Life in the Time of Slow Hauling Knowledge

Dolores Steinman

As 2020 began, the world faced an invisible, yet palpable presence: the novel virus, SARS-CoV-2. Connections with colleagues online enabled a form of citizen science in the pandemic's early days, sharing knowledge and tentative research on a small scale, reflecting larger initiatives globally. We each absorbed the information and the RNA differently. We each reacted differently both to the situation as well as the virus. Recounting my journey as one of COVID-19's earliest patients, near the European epicenter in Bergamo, Italy, in 2020, the following text is more companion to discovery than *vade mecum*. Tracing my personal experience of long COVID, as a trained pediatrician and cell biologist, with experience in the arts, it can be read in the context of narrative medicine (Charon 2001). I document my stages of experiencing the virus (both physical and political) and the pandemic's social consequences during this time.

This is not a story of loss or grief. Unless we think of loss of trust, or grieving our spent hope. I would rather think of it as a story of discovery in seven stages: shock, denial, pain, guilt, anger, depression, and acceptance. The way I experienced it, nonetheless, the stages seemed to be: anger, frustration, frustration, frustration, despair, relief, forgetting. When later asked, "How did you fare?", all the fears, and bile, pain, and sorrow came pouring out furiously.

After news of a novel coronavirus outbreak in Wuhan, I prepared to return from my workplace in Canada to my home away from home in Italy. I happened to land in Milan on February 17, 2020, and traveled to our home no more than 20 kilometers from Bergamo, which we later discovered was the pandemic's epicenter at that time.

Leaving Toronto, I was masked; once out of the Milan terminal, no such restraints applied. Day after day of not-so-blissful ignorance: a big international

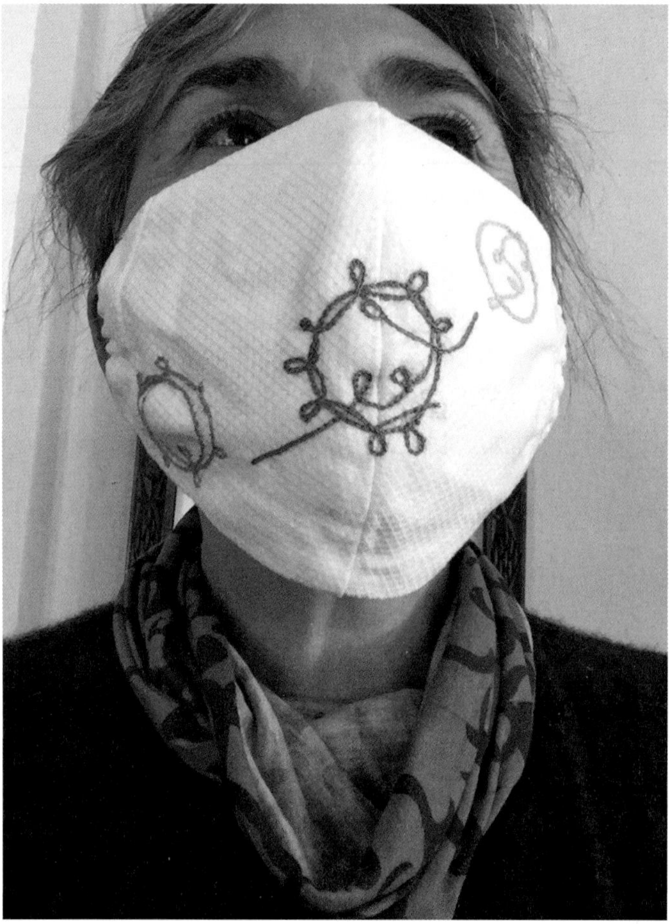

FIGURE 15.1 *Dolores Steinman,* Embroidered Imaginarium, *2020. Cotton Embroidery on Home-made Cotton Piquet Mask. Courtesy of Dolores Steinman.*

soccer game, the weekly itinerant open market, friends and neighbors meeting for coffee or aperitivo, grandparents walking little ones down the riverbank. After adhering to COVID restrictions set on March 11, I did not socialize, or even step out, other than for necessary groceries. However, an unfortunate coffee gaffe, which started cautiously and innocently, ended with an ill-omened hug.

My friend and neighbor lost her husband to COVID-19. She recovered but will never mend her broken life. A couple of weeks earlier, I had invited her to my apartment for coffee and hugged her, exposing me to COVID-19. This one visit altered the rest of my spring, summer, and maybe the rest of my days.

Today, I may still harbor viral remnants, as virology professor Marc-André Langlois hypothesized (Molina 2020), but I certainly carry the sequelae, or after-effects.

With testing not yet commonplace, my illness was undefined. Symptoms I originally blamed on prolonged "Zooming" turned out to be manifestations of the virus itself. Little did we know, we had so much hope: my "first day" coincided with the day the United States announced the first human trials of an anti-CoV-2 vaccine (Watts, Jackson, and Maxouris 2020).

Church bells in my neighborhood began tolling several times daily. I rested in my apartment, blinds drawn to keep light from hurting my tired eyes. My journey through COVID-19 consisted of waves of weakness and floods of symptoms alternating with waves of calm and relative wellbeing (albeit with overwhelming fatigue). They were almost in sync with waves of information regarding new outbreaks. Slowly, it became clear that the original two-week time course was, in fact, very elastic: some of us would glide through, while others crawled to the finish line; some would never make it. The sprint was turning into a marathon.

Why weren't the doctors—who advised my friends and neighbors to take fever-suppressing antipyretics and carry on—admitting their mistake and preventing the virus's spread? Why were they letting the population still believe that minor symptoms were not part of the COVID-19 description, and allowing "fatigued" people to walk freely? Where did common sense go?

I remember my reaction when asked why I was wearing a mask. I proposed that this person conduct an experiment. I instructed:

Hold your breath.
Put a bit of talcum powder in your nostrils.
Breathe out (not sneezing necessarily or blowing hard).
Watch and see how far it goes.
Calculate the size and weight of the powder "particle" and compare with tiny droplets carrying coronavirus.
Calculate the approximate distance to where the droplet hits a surface, where it will reside for several hours.

I was no Richard Feynman (the physicist famed for his ability to translate complex information into simple experiments). However, I wondered why there were no clear explanations about the virus once information was available (Matheson and Lehner 2020).

In the following weeks, waves of confusion and frustration broke with the same strength and rhythm. There were waves of relative comfort followed by lassitude, malaise, pain, and weakness. The dry throat, unbearable headache, paralyzing muscular pains, vivid nightmares-turned-hallucinations, my skin burning from within. My days were shortened by lethargy, dragging away my confidence and faith in my own body. Fighting fever is easier than fighting weariness. Particularly when available information is confusing, rushed, and inaccurate.

Most astonishing to me was the "shooting from the hip," the hurry to come up with something, anything, a hollow promise. Many clinicians publicized, and sometimes even published, truncated ideas and plans for treatments and vaccines, without sharing or explaining crucial details. Trained to never give unrealistic expectations or create confusion, as a patient it infuriated me to become buried under the avalanche of media sound bites. The summary of the condition was confusion, as Ed Yong stated regarding immunology's complexity: "Bad decisions mean more cases, which means a wider variety of possible immune responses, which means a higher prevalence of rare events. In other words, the worse the pandemic gets, the weirder it will get" (2020).

Did any of the physicians, giving false hope with promises of quick solutions or disregarding "minor" symptoms, think how dangerous that was, how betrayed the patient felt, how much pain they caused? When approached by patients—long after the main symptoms receded—would they understand how demeaning it was to disregard their experiences? How many of us are still aching, unable to focus, weak or walking through fog?

Later, I learned I wasn't alone: "One of the most common things you'll hear from long-haulers is that doctors have told them repeatedly that their symptoms are just anxiety or stress or in their heads. And to be clear, these are people who have crushing fatigue. They have all these incredibly intense physical symptoms, and they are being turned away by people who just don't believe them" (Mimbs Nyce 2020). As discussed within my art-science community online, was there an imbalance in who was being disregarded? It was not long before evidence began to suggest so (see Chapters 14 and 18 in this volume).

By now, hospitals overflowed, military trucks carried out the dead, and church bells stopped tolling. With helplines busy and hospitals overloaded, there was no access to PCR tests, unless you were at death's door. After six weeks of relentless illness, I panicked. Breathing was not getting easier, pain more tolerable, or sleep more restful. I listed everything I still wanted to tell my daughter if I did not make it. Weeks later, when antibody testing became available, my weakened system revealed my antibody titer was low. I felt cheated. After all the suffering, no safety, no hope?

Meanwhile, under immense time pressure and a surging death toll, frightened scientists ran in all directions: antibody treatment, vaccine, symptomatic treatment? Scientists knew the genetic sequence, but nobody could agree on the sequence of events. As antibodies became the focus—generating them as well as isolating from recovered patients—the question became: will they be long-lived and reliable?

One can think of the pandemic as an incredible opportunity for knowledge production. As months progressed, so did our understanding of the process we were undergoing. After the first moment of shock, to keep our minds occupied, challenges sprouted on the internet: poetry, culinary recipes, lists of one's top ten books, movies, artworks. And then, this effort of being together despite isolation, of keeping informed, stimulated, and focused, generated online initiatives like Viral Cultures (2020), FEMeeting Teapot Chat (2020), and Art Resilience Economy (ARE, 2020), bringing our large art-science family together, in an attempt to crystallize or distill all fleeting thoughts and volatile feelings. I am grateful to these online gatherings' organizers Claire Nettleton (Viral Cultures), Dalila Honorato and Marta de Menezes (FEMeeting TeapotChat), and Anna Walker (ARE)—their efforts were the buds of a current "Medici effect." This structured, yet boundary-free, convergence of minds made it possible not only to rely on each other for support but also to see this new world through different lenses, addressing issues of universal concern.

On a smaller scale, we reflected the larger initiatives observed globally, that made new, targeted research possible, as well as raising awareness of COVID-19's long-term consequences (Basu 2020) before the established laboratories confirmed them (Schurink et al. 2020).

This experience reminded me of the porosity and permeability of our bodies and our cultures and how our environments—social, political, and economic—molded our response to the pandemic (Meloni 2018). We each tried, in our own way, to cope with the change, adjust to it, and alleviate the pain—our own and others'. Physical reality is not, after all, disconnected from our minds (see Chapter 21 in this volume). In what might be considered a Dada-esque attempt to come to terms with the virus, I have counted the number of bananas consumed during confinement weeks, made cheese-rind poetry, and embroidered everything with my interpretation of a SARS-CoV-2 virus (Figure 15.1).

As we move out of the pandemic's grip, we should not approach our ongoing interaction with the virus as a confrontation with evil but look at ways in which we will learn to coexist (Peiris and Leung 2020). We should neither deny nor demean it, ignore or fear it, but try to observe, reflect, and learn. Studies show that in death, men were struck more, whereas in life women were affected more (Collins 2020). Is there a chance to correct such discrepancies? Can we give disadvantaged groups in society a say and a way? There is no "silver bullet," as we are individuals (Mathew et al. 2020), yet I believe "being in it together" (Braidotti 2020) is the best and most sustainable option.

References

Basu, T. (2020). "Covid-19 'Long Haulers' are Organizing Online to Study Themselves." *MIT Technology Review*. https://www.technologyreview.com/2020/08/12/1006602/covid-19-long-haulers-are-organizing-online-to-study-themselves/

Braidotti, R. (2020). "'We' Are In This Together, But We Are Not One and the Same." *Bioethical Inquiry* 17(4): 465–9. doi:10.1007/s11673-020-10017-8

Collins, C. (2020). "Productivity in a Pandemic." *Science* 369(6504): 603. doi: 10.1126/science.abe1163

Matheson, N. J., and Lehner, P. J. (2020). "How Does SARS-COV-2 Cause COVID-19?" *Science* 369(6503): 510–11. doi:10.1126/science.abc6156

Mathew, D., et al. (2020). "Deep Immune Profiling of COVID-19 Patients Reveals Distinct Immunotypes with Therapeutic Implications." *Science* 369(6508). doi: 10.1126/science.abc8511

Meloni, M. (2018). "A Postgenomic Body: Histories, Genealogy, Politics." *Body & Society* 24(3): 3–38. doi: 10.1177/1357034X187854

Mimbs Nyce, C. (2020). "A Q&A With Ed Yong on Long-Haulers." *The Atlantic*. https://www.theatlantic.com/newsletters/archive/2020/08/ed-yong-long-haulers-covid/615430/

Molina, K. (2020). "Lingering Symptoms among COVID-19 'Long-haulers' Remain a Mystery." *CBC News*. https://www.cbc.ca/news/canada/ottawa/covid-long-haulers-still-suffering-months-later-1.5688673

Peiris, M., and Leung, G. M. (2020). "What Can We Expect from First-generation COVID-19 Vaccines?" *The Lancet* 396(10261): 1467–9. doi: 10.1016/S0140-6736(20)31976-0

Schurink, B., et al. (2020). "Viral Presence and Immunopathology in Patients with Lethal COVID-19: A Prospective Autopsy Cohort Study." *The Lancet Microbe* I(7). doi:10.1016/s2666-5247(20)30144-0

Watts, A., Jackson, A., and Maxouris, C. (2020). "Why These Volunteers Chose to Participate in a Coronavirus Vaccine Trial." *CNN*. https://edition.cnn.com/2020/03/17/health/coronavirus-trial-vaccine-neal-browning-trnd/index.html

Yong, E. (2020). "Immunology is Where Intuition Goes to Die." *The Atlantic*. https://www.theatlantic.com/health/archive/2020/08/covid-19-immunity-is-the-pandemics-central-mystery/614956/

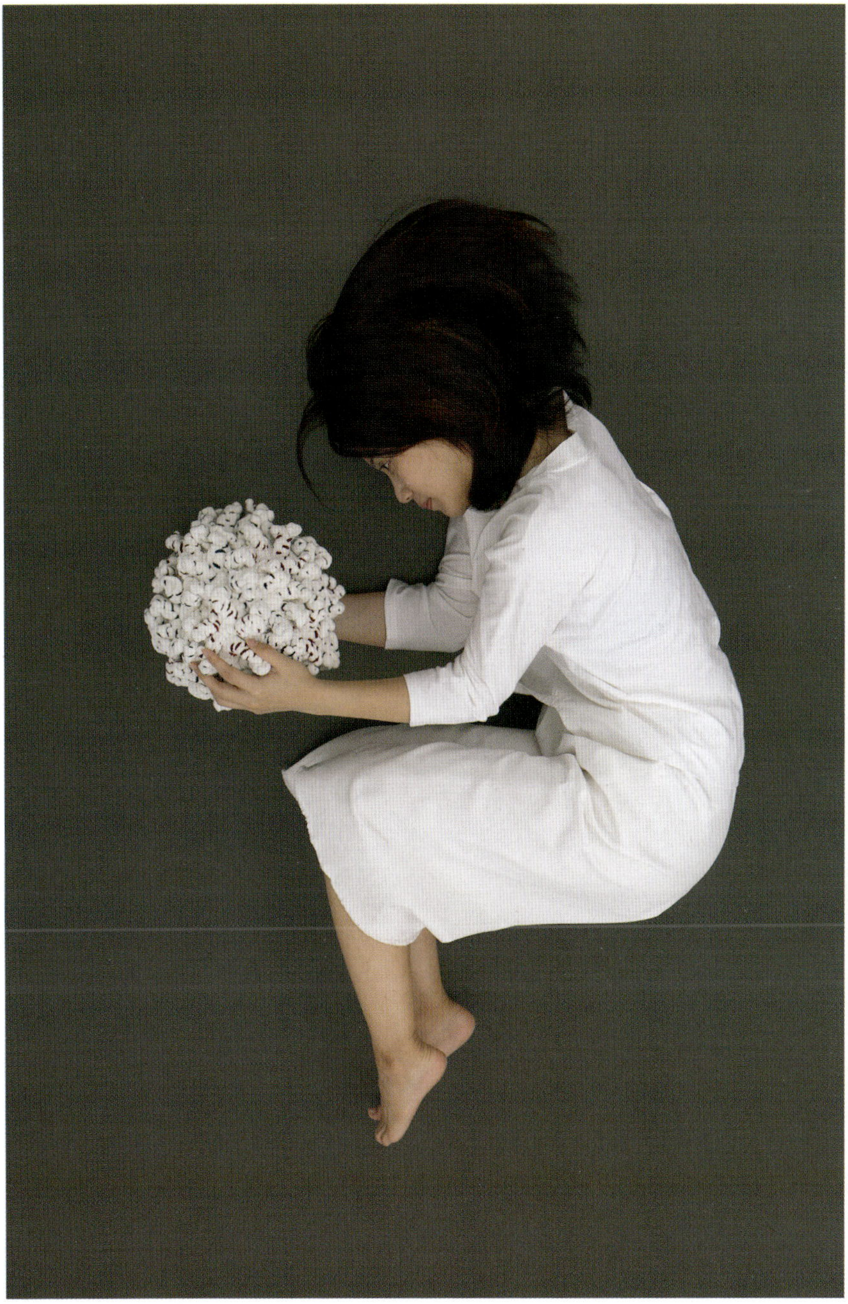

PLATE 1 Pei-Ying Lin, *Tame is to Tame*, 2016. Courtesy of the artist

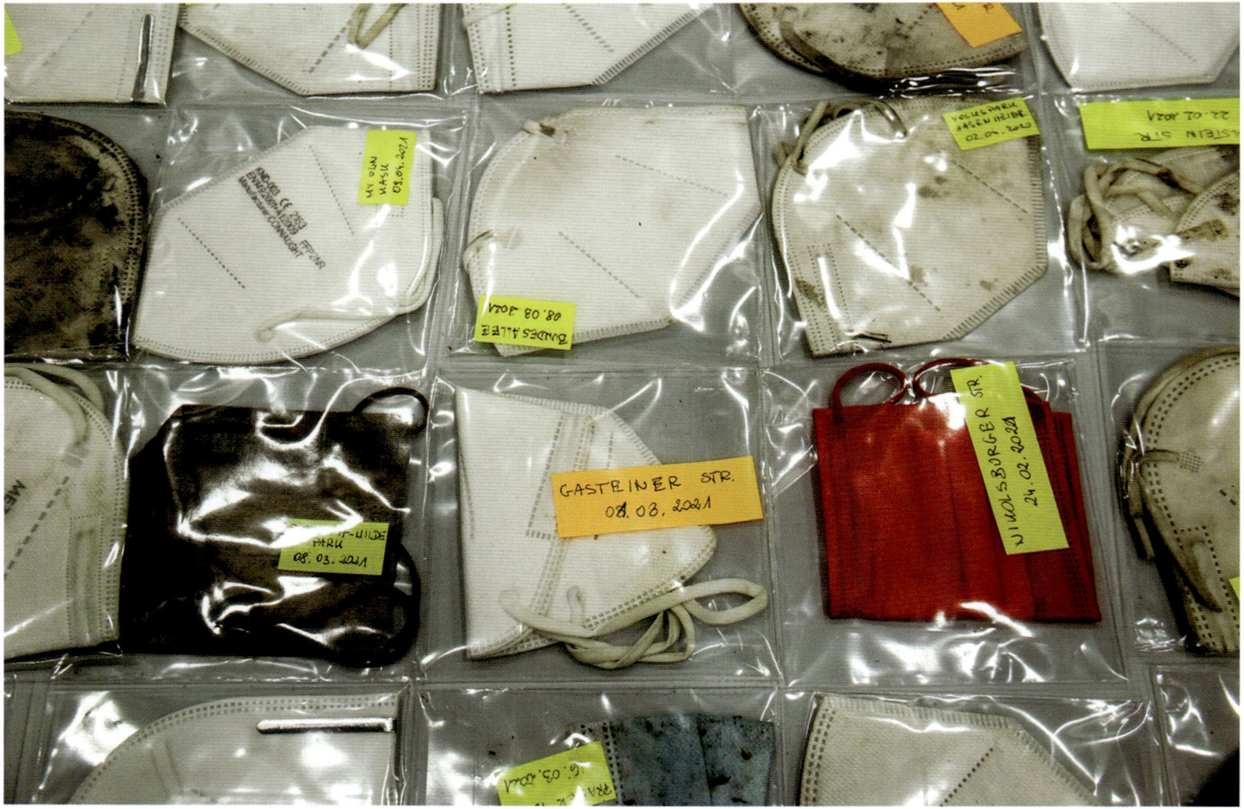

PLATE 2 Karolina Zyniewicz, *Sign of the Times*, 2021. Image: Adam Bogdan. Courtesy of the artist

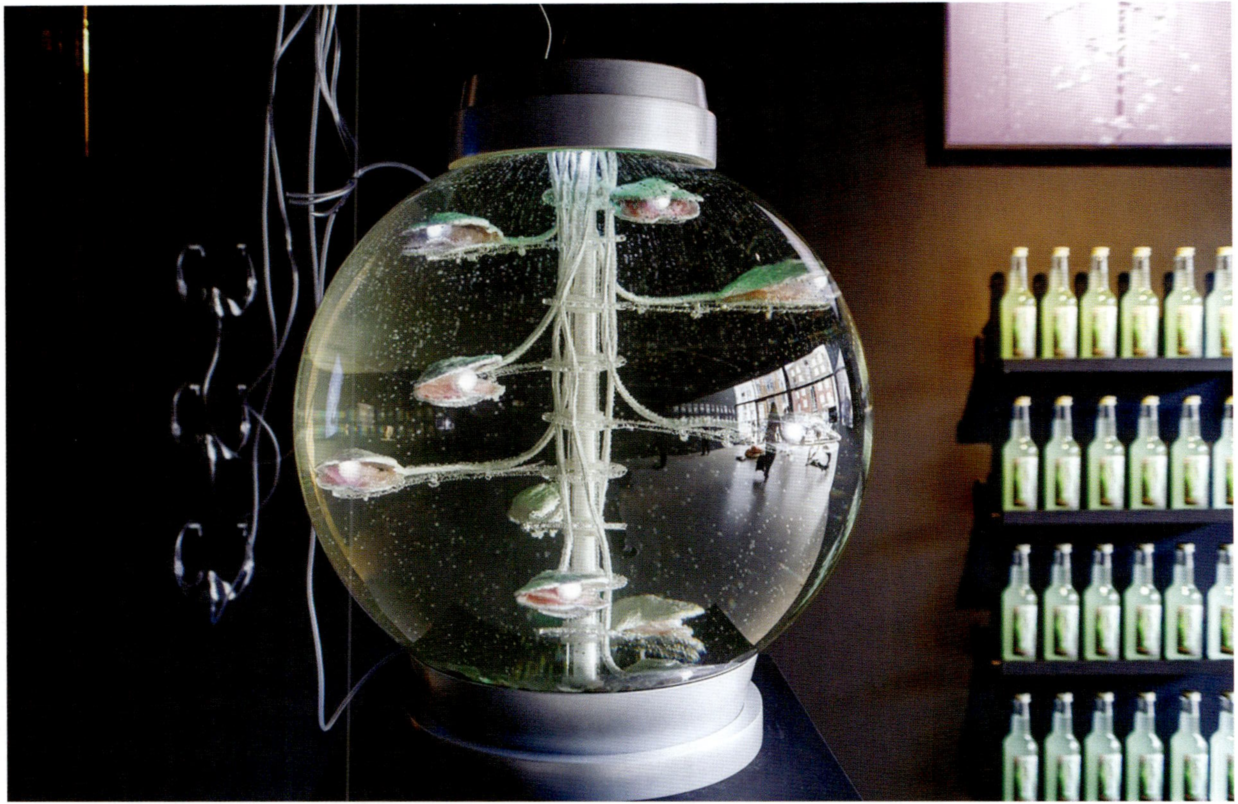

PLATE 3 Stephanie Rothenberg, *Aquadisia*, 2022. Courtesy of the artist

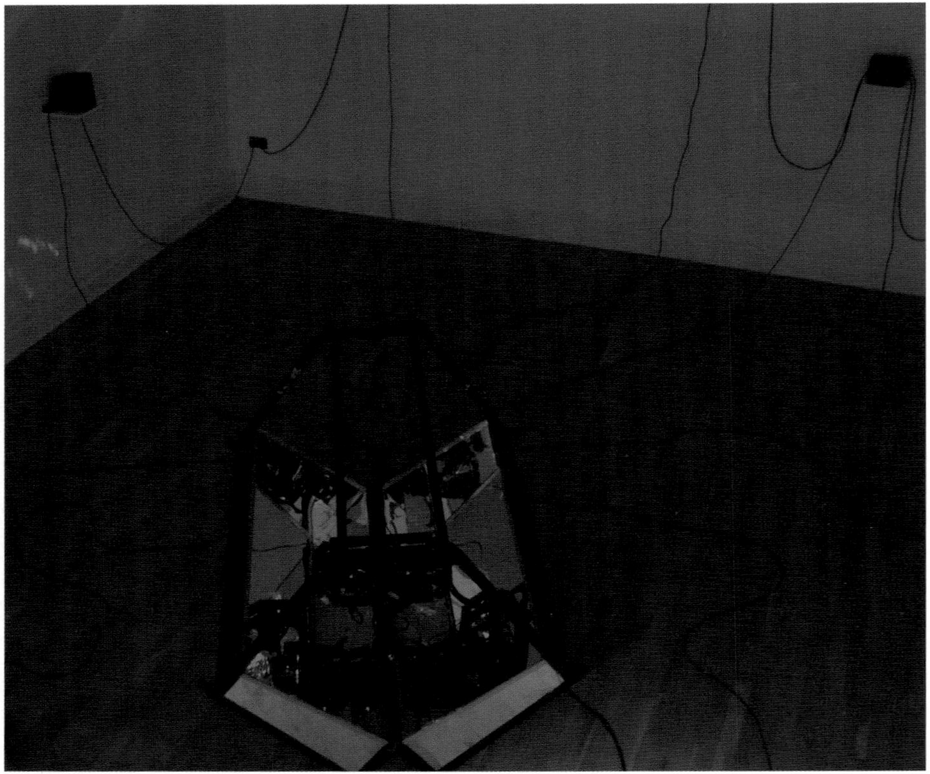

PLATE 4 Louise Mackenzie, *-Phage*, 2017. Installation View, Queen's Hall Arts, Hexham. Courtesy of the artist

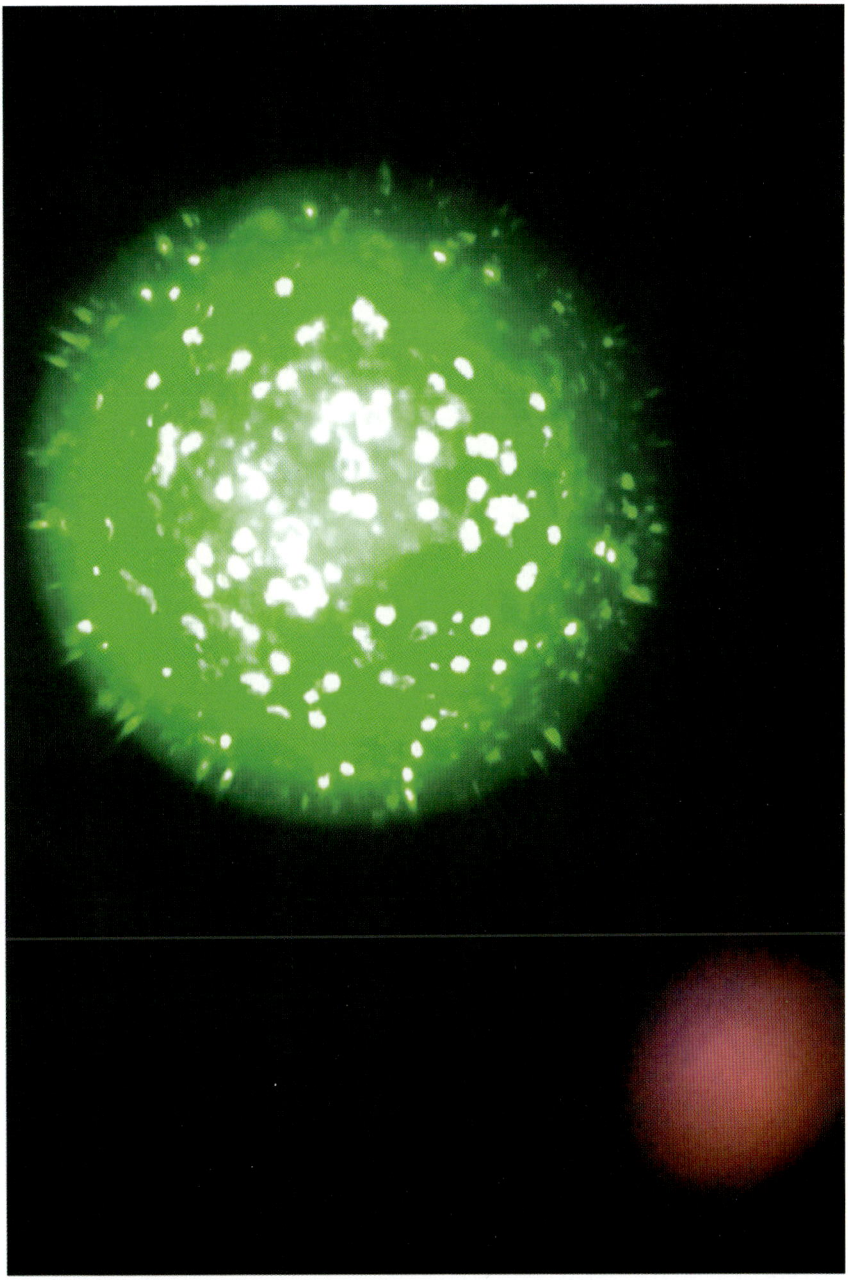

PLATE 5 Louise Mackenzie, *Blue Green Planet*, 2013. C-Type Print. © Louise Mackenzie, 2013

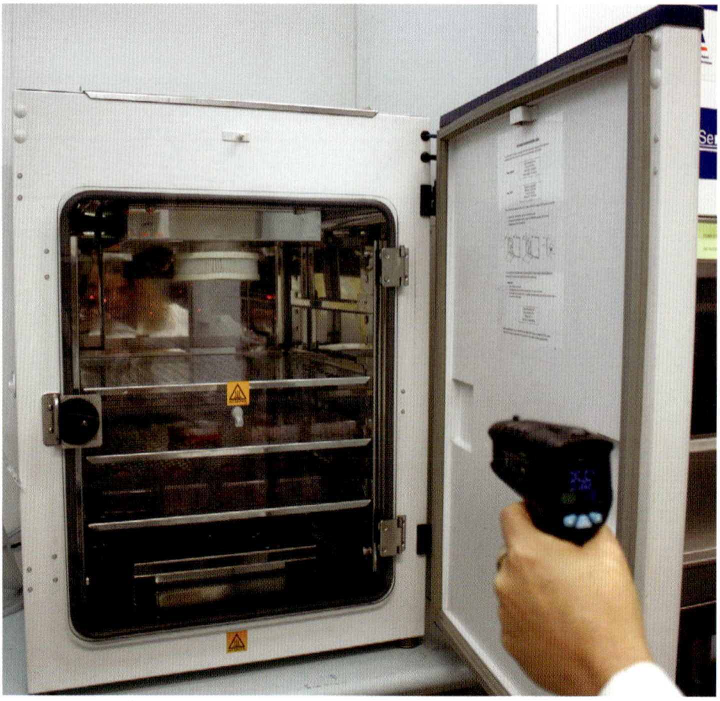

PLATE 6 Oron Catts and Ionat Zurr, *Thermoception*, 2020. From Sensory Orders Exhibition, ŁAŹNIA Centre for Contemporary Art (PL). Courtesy of the artists

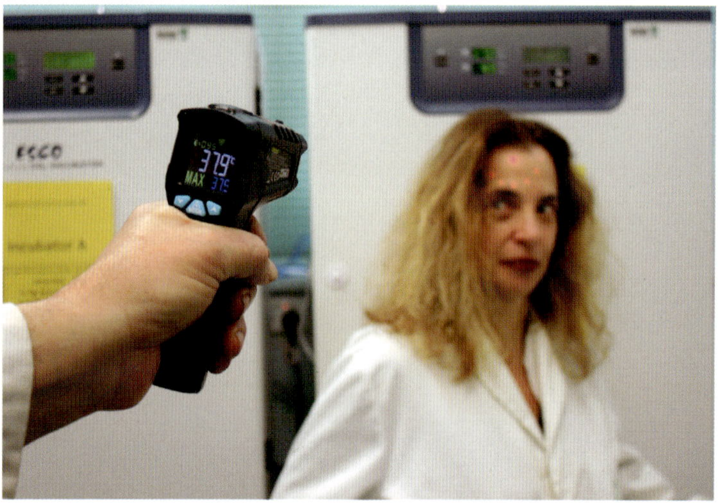

PLATE 7 Oron Catts and Ionat Zurr, *Thermoception*, 2020. From Sensory Orders Exhibition, ŁAŹNIA Centre for Contemporary Art (PL). Courtesy of the artists

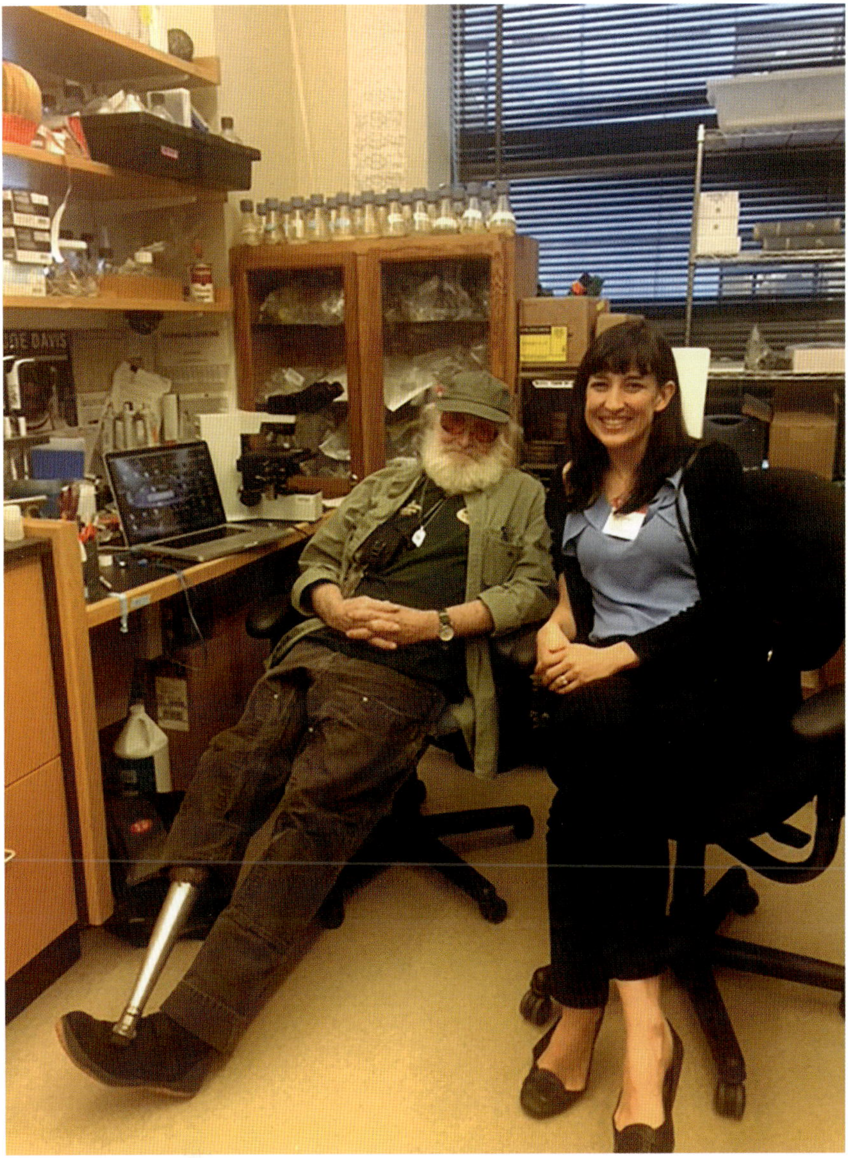

PLATE 8 Claire Correo Nettleton and Joe Davis at Church Lab, 2018. Courtesy of the author

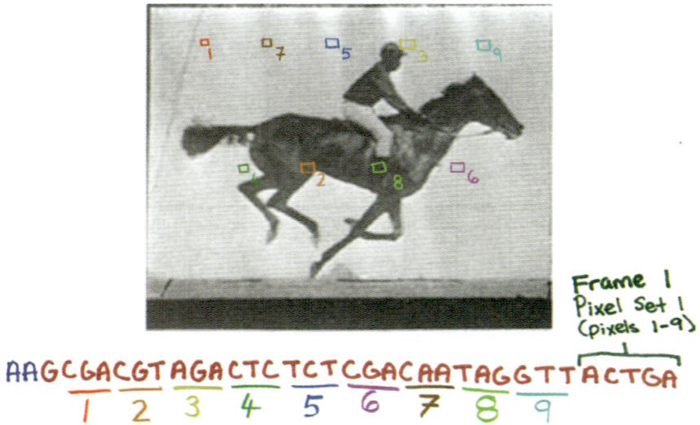

PLATE 9 *Encoding Memories in Living Cells with CRISPR*, 2017. Courtesy of the Wyss Institute

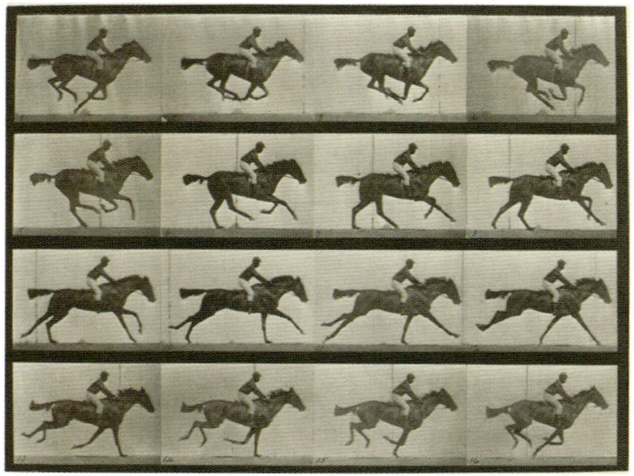

PLATE 10 Edweard Muybridge, *Animal Locomotion, Plate 626*. Photomechanical Print: Collotype, 1887, Boston Public Library

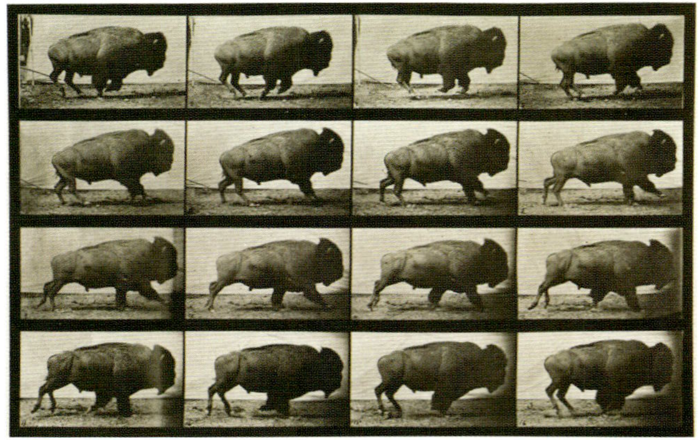

PLATE 11 Edweard Muybridge, *Animal Locomotion. Buffalo; galloping: an electro-photographic investigation of consecutive phases of animal movements, Plate 700*. Photomechanical Print: Collotype, 1887, Boston Public Library

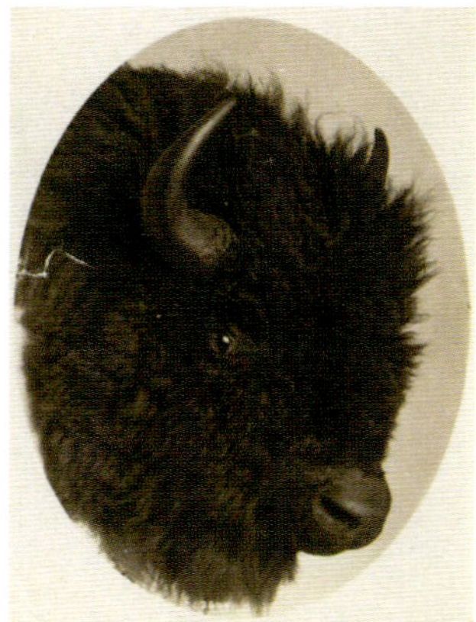

PLATE 12 William Henry Jackson, *Bull*, Albumen print, 1875, The Benton Museum of Art at Pomona College. Gift of James E. and Debra Ann Pearl

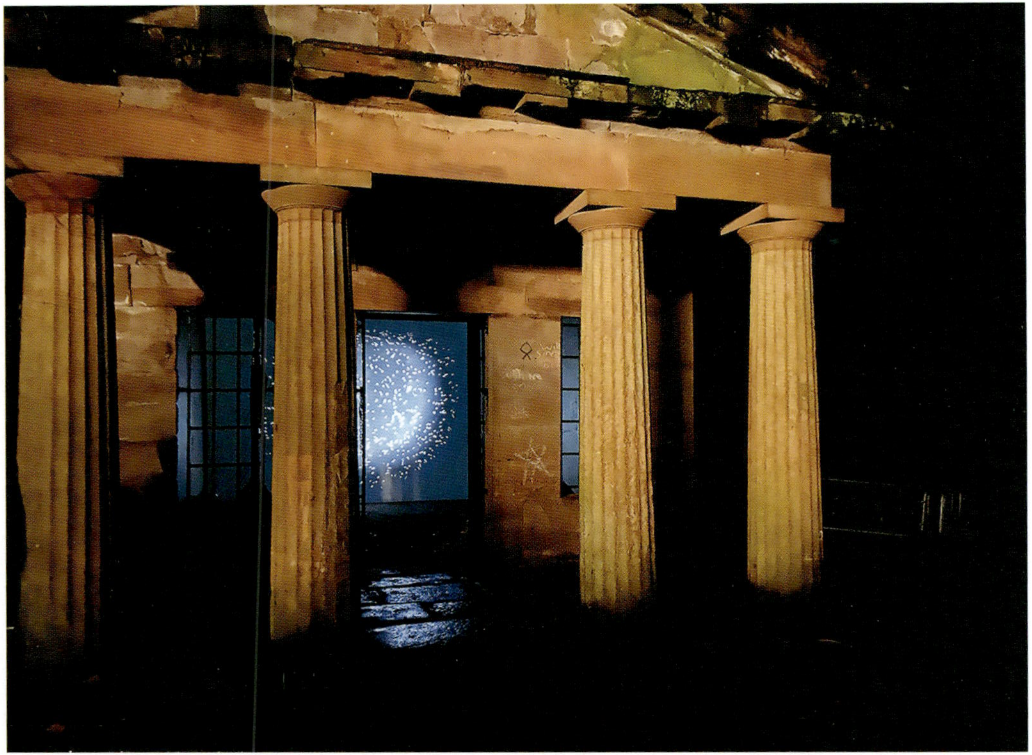

PLATE 13 Louise Mackenzie, *The Stars Beneath Our Feet*, 2015. Site-sensitive audio-visual installation, 7:07 (loop). Commissioned for Lumiere Durham 2015, a four-day international light festival produced by Artichoke in the UK. Courtesy of the artist

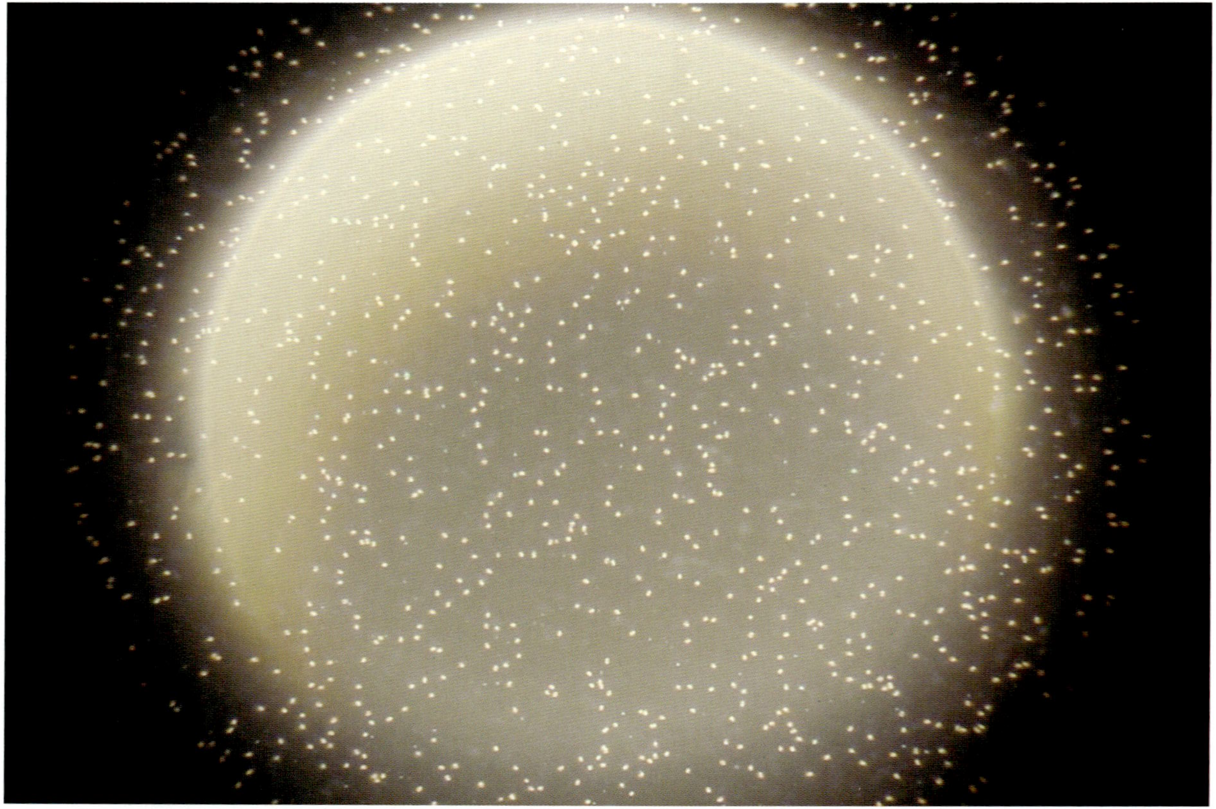

PLATE 14 Louise Mackenzie, *The Stars Beneath Our Feet*, 2015. Video still. © Louise Mackenzie, 2015

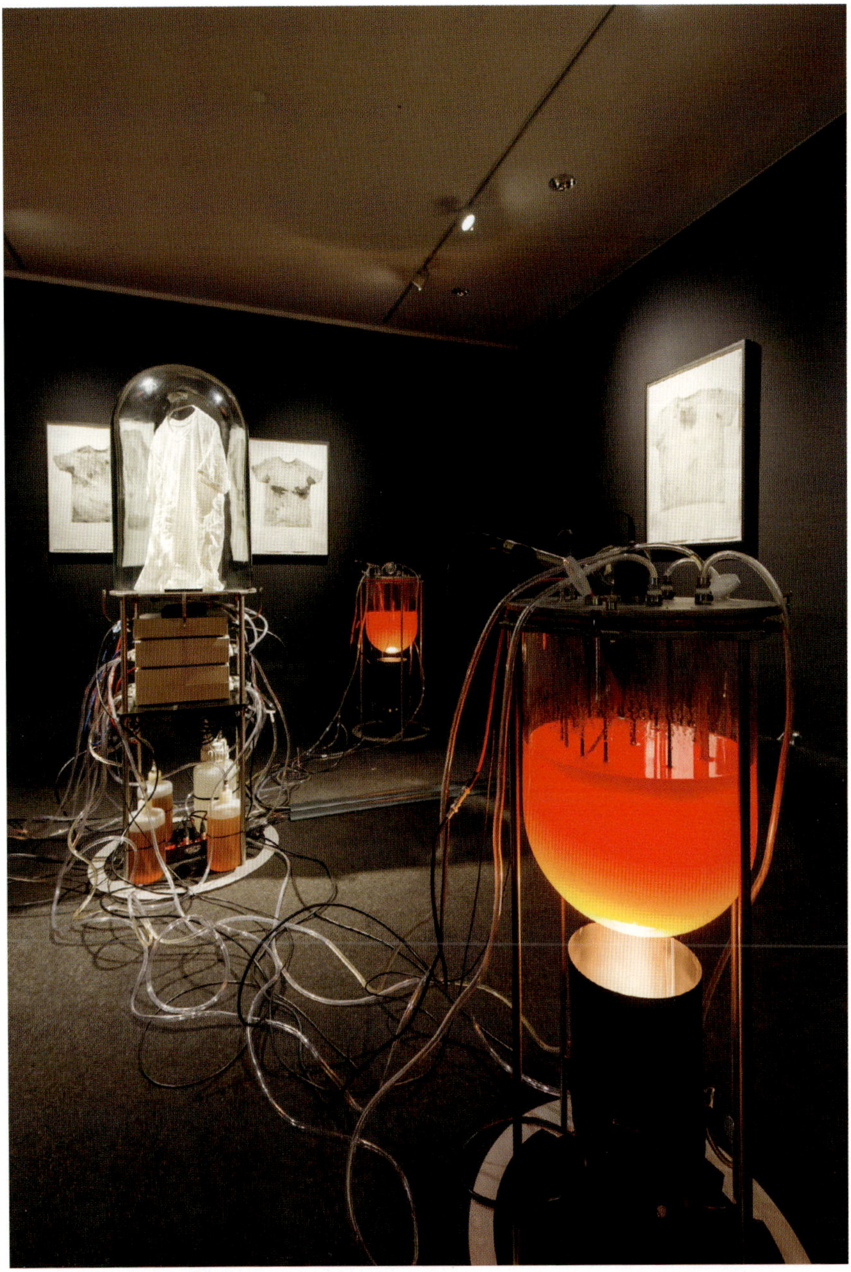

PLATE 15 Paul Vanouse, *Labor,* Multi-sensory installation, Prix Ars Electronica Festival, OK Center, Linz, Austria, 2019. Installation shot: CYBERARTS OÖ Kulturquartier, Linz. Photo by Otto Saxinger. Image courtesy of the artist

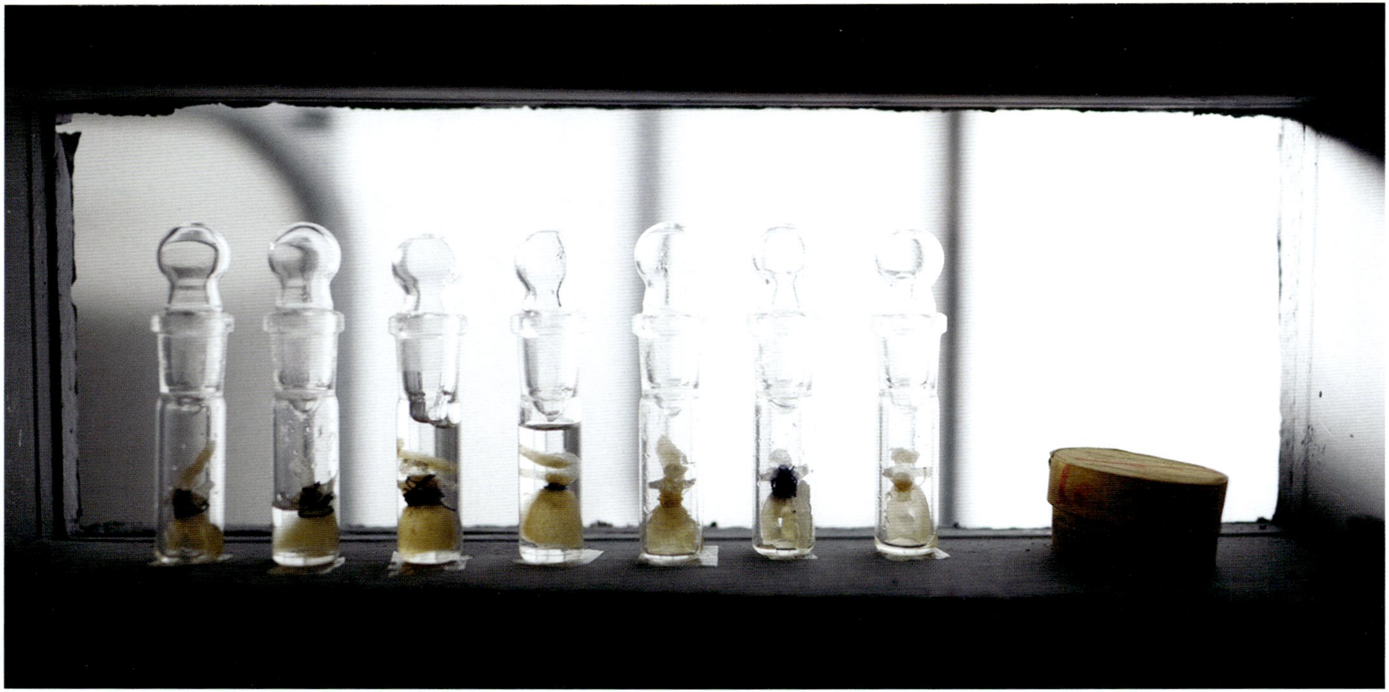

PLATE 16 Oron Catts and Ionat Zurr (The Tissue Culture & Art Project), *The Semi-Living Worry Dolls*, Cells, biodegradable/bioabsorbable polymers and surgical sutures, 2000. Courtesy of The Tissue Culture & Art Project (Oron Catts & Ionat Zurr)

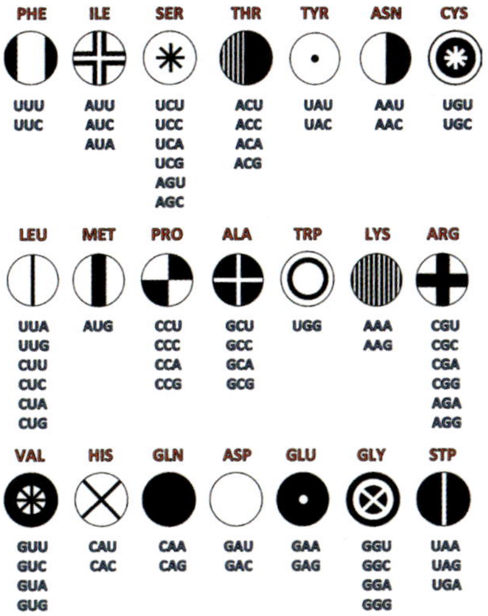

PLATE 17 Joe Davis. *Graphical Representation of 64 codons and corresponding amino acids (genetic code)*, 2021. Courtesy of Joe Davis

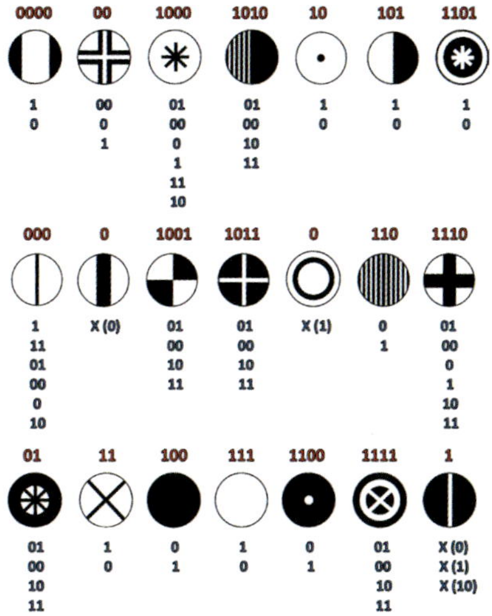

PLATE 18 Joe Davis. *Graphical Representation of Amino Acids which have been assigned distinct values* (amino code), 2021. Courtesy of Joe Davis

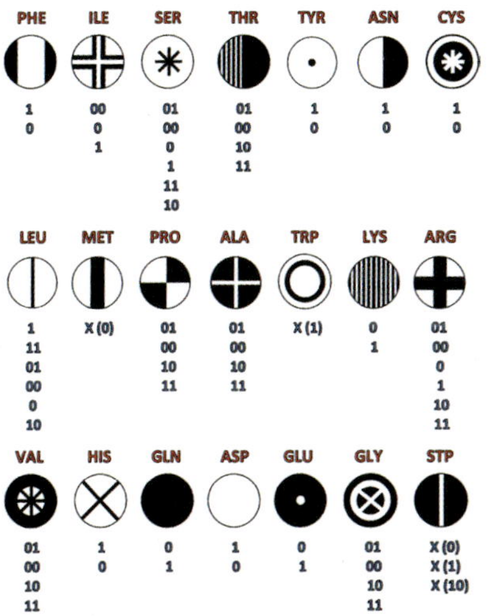

PLATE 19 Joe Davis. *Graphical Representation of Amino Acids with Values Assigned According to Mass* (silent code), 2021. Courtesy of Joe Davis

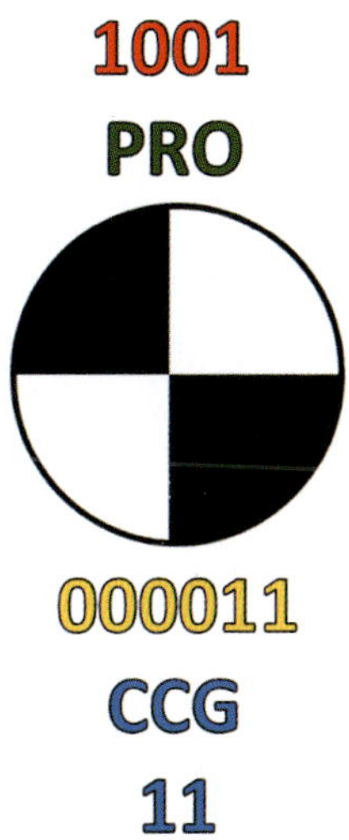

PLATE 20 Joe Davis. *Three numbers embedded within DNA Molecules: Amino (1001), Codon (11) and DNA (000011)*, 2021. Courtesy of Joe Davis

SUBHAN ALLAH MANIFOLD ENCODING:

Amino Code:	1101	1000	1011	00	1111	01	1000	1010	1000	110	1100	01	01	01	1011
Amino acid:	CYS	SER	ALA	ILE	GLY	VAL	SER	THR	SER	LYS	GLU	VAL	VAL	VAL	ALA
DNA:	TGT	TCG	GCT	ATA	GGC	GTT	TCT	ACA	TCT	AAG	GAG	GTA	GTG	GTC	GCT
Silent Code:	1	1	01	1	00	01	01	10	01	1	1	10	11	00	01

Amino Code:	1011	000	1010	01	111	1011	00	101	1010	000	01	000	00	1101	1000	1010
Amino acid:	ALA	LEU	SER	VAL	ASP	CYS	ILE	ASN	THR	LEU	VAL	LEU	ILE	CYS	SER	THR
DNA:	GCT	CTT	TCC	GTT	GAT	TGC	ATA	AAT	ACC	CTT	GTT	CTT	ATA	TGC	AGT	ACA
Silent Code:	01	01	00	01	1	0	1	1	00	01	01	01	1	0	11	10

Amino Code:	01	11	11	01	1001	1000	01	00	11	01	1001	1000	01	00	1101	1001
Amino acid:	VAL	HIS	HIS	VAL	PRO	SER	VAL	ILE	HIS	VAL	PRO	SER	VAL	ILE	CYS	PRO
DNA:	GTG	CAC	CAC	GTT	CCT	TCC	GTG	ATA	CAT	GTA	CCG	TCC	GTA	ATA	TGT	CCT
Silent Code:	11	0	0	01	01	00	11	1	1	10	11	00	10	1	1	01

Amino Code:	1000	01	11	–	111	01	110	1110	1110	1110	1110	1110	1110	1110	1110
Amino acid:	SER	VAL	HIS	–	ASP	VAL	LYS	ARG	ARG	ARG	ARG	ARG	ARG	ARG	ARG
DNA:	TCC	GTC	CAC	–	GAT	GTC	AAA	CGC	AGG	CGT	AGA	CGC	AGA	AGA	CGT
Silent Code:	00	00	0	–	1	00	0	00	11	01	10	00	10	10	01

Amino Code:	1110	1110	1110	1110	1110	1110	1110	1110	1110	1110	1110	1110	1110	1110
Amino acid:	ARG	ARG	ARG	ARG	ARG	ARG	ARG	ARG	ARG	ARG	ARG	ARG	ARG	ARG
DNA:	AGG	AGG	CGT	AGA	CGT	AGA	CGC	CGT	CGC	AGG	CGT	AGA	CGT	AGA
Silent Code:	11	11	01	10	01	10	00	01	00	11	01	10	01	10

Amino Code:	1110	1110	1110	1110	1110	1110	1110	1110	1110	1110	1110
Amino acid:	ARG	ARG	ARG	ARG	ARG	ARG	ARG	ARG	ARG	ARG	ARG
DNA:	CGC	CGT	CGC	AGG	CGT	AGA	CGT	AGA	CGC	CGT	AGG
Silent Code:	00	01	00	11	01	10	01	10	00	01	11

PLATE 21 Joe Davis, *Subhan Allah DNA Manifold*, 2021. Courtesy of Joe Davis

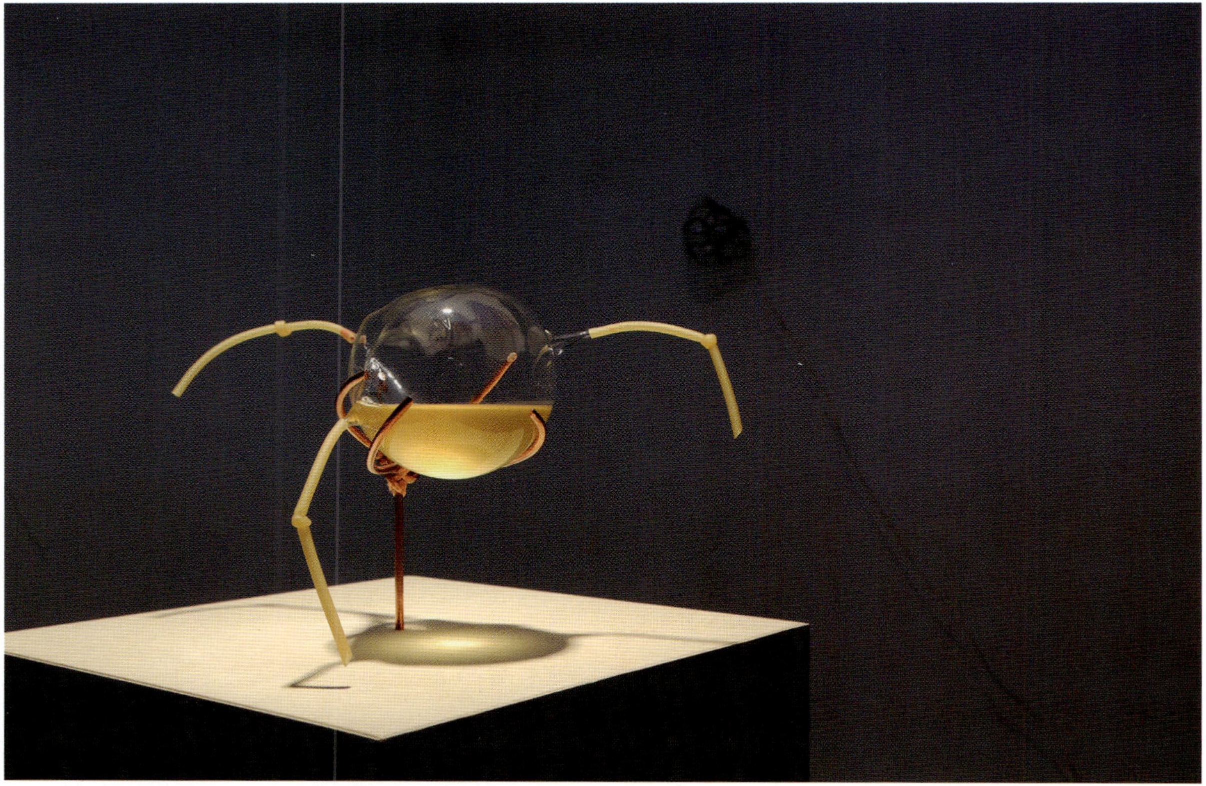

PLATE 22 Louise Mackenzie, *Pithos ConsTrained*, 2018. Installation View, Gallery North Project Space, Newcastle upon Tyne, UK. Image: Jason Revell. © Louise Mackenzie

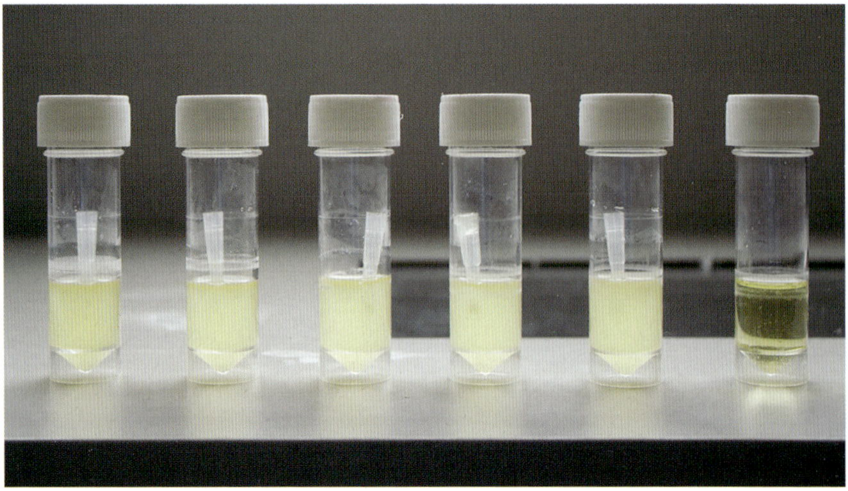

PLATE 23 Louise Mackenzie, *Lively Material*, 2018. Video Still. © Louise Mackenzie

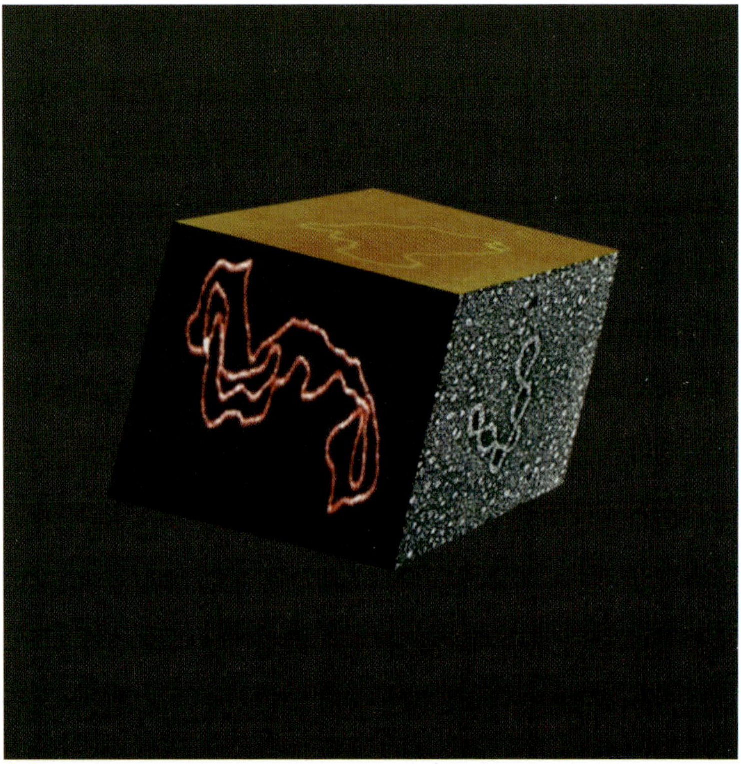

PLATE 24 Louise Mackenzie, *Flat Lively Objects*, 2017. C-Type Print. © Louise Mackenzie

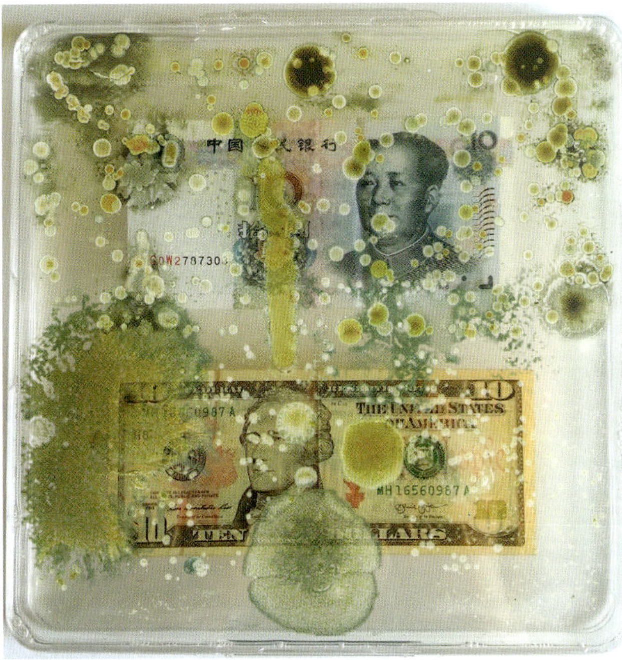

PLATE 25 Ken Rinaldo, *Borderless Bacteria / Colonialist Cash*. Installation detail. Photo: Art Laboratory Berlin, January 2020. Courtesy of Art Laboratory Berlin

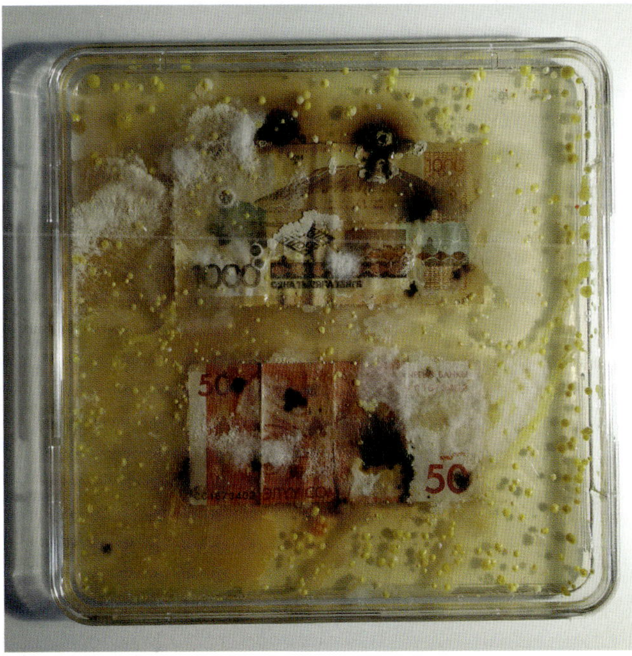

PLATE 26 Ken Rinaldo, *Borderless Bacteria / Colonialist Cash*. Installation detail. Photo: Art Laboratory Berlin, January 2020. Courtesy of Art Laboratory Berlin

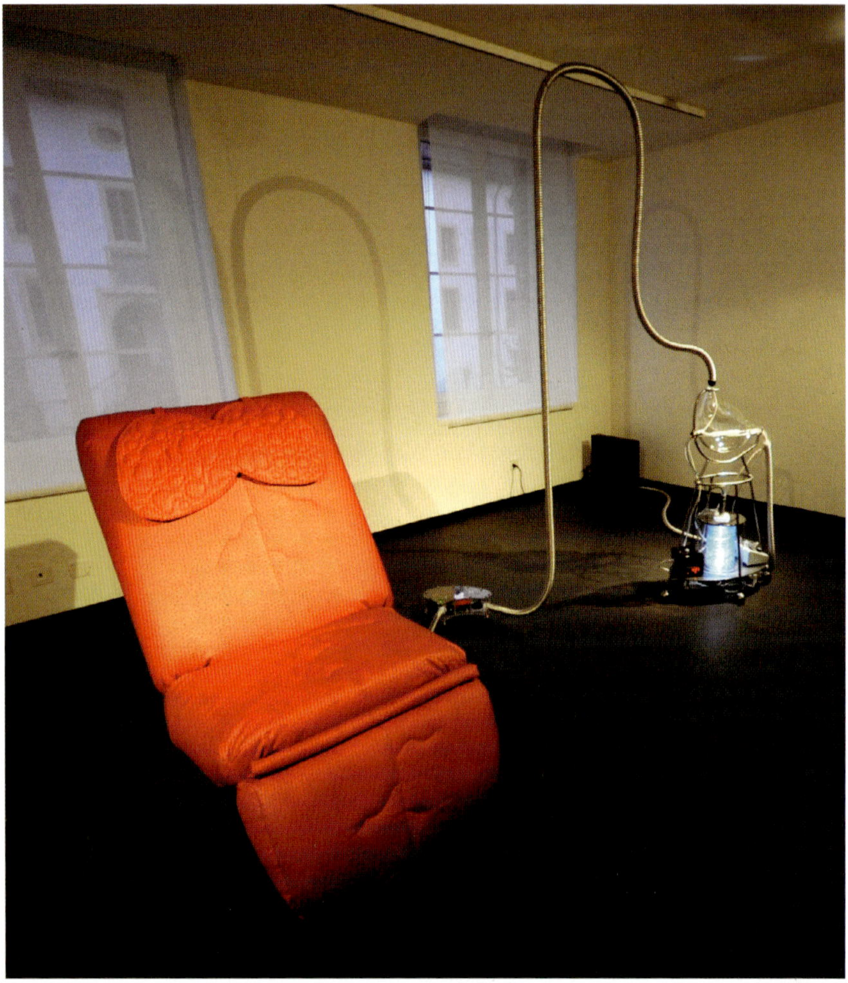

PLATE 27 Ken Rinaldo, *The Enteric Consciousness*. Maison d'Ailleurs Museum, Switzerland, 2010. © Ken Rinaldo. Photo: Joanna Avril

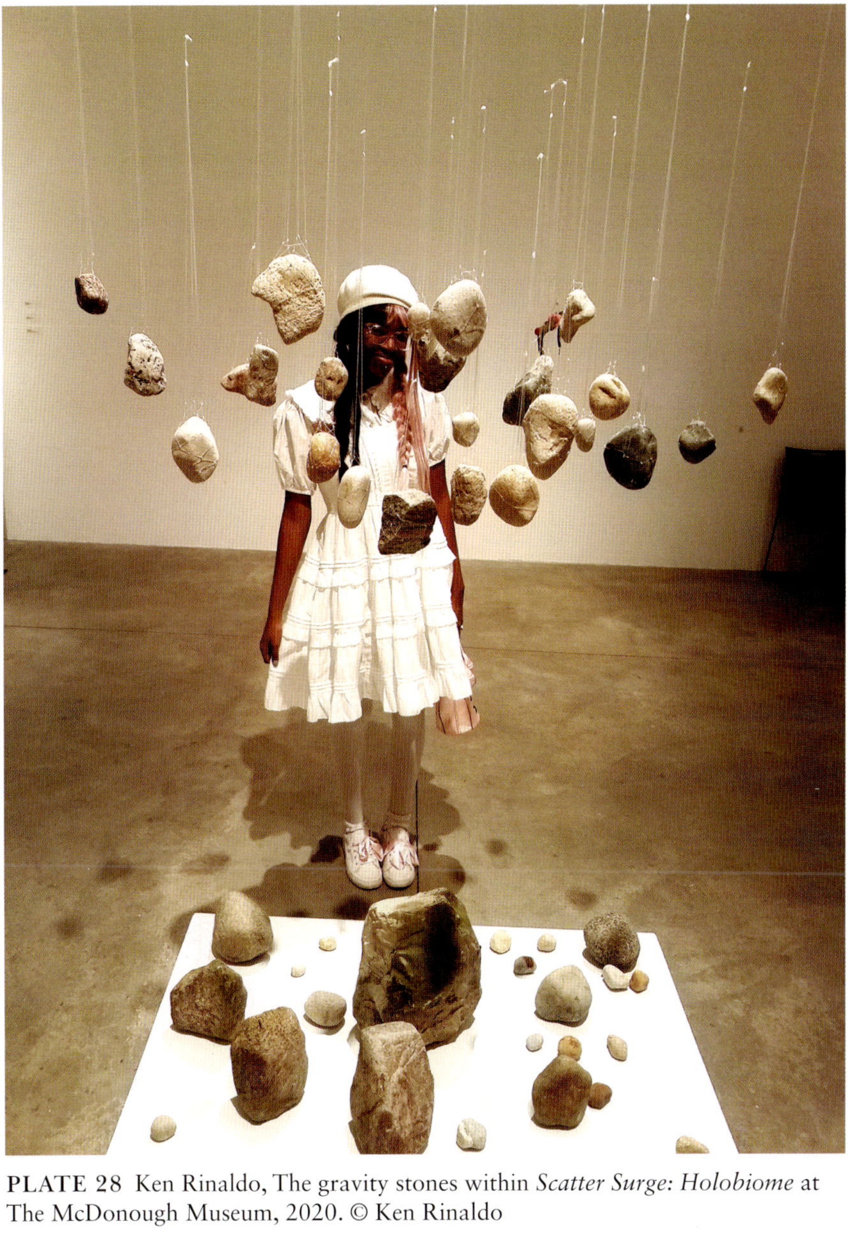

PLATE 28 Ken Rinaldo, The gravity stones within *Scatter Surge: Holobiome* at The McDonough Museum, 2020. © Ken Rinaldo

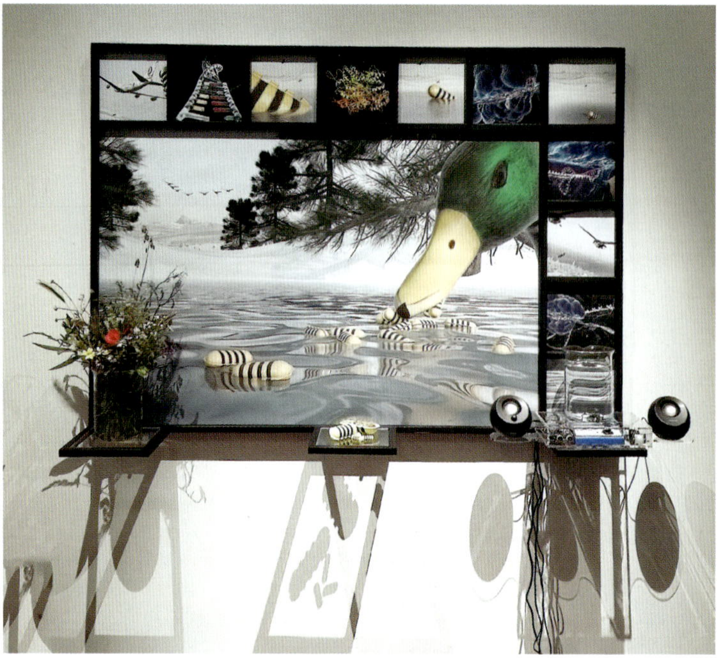

PLATE 29 Ken Rinaldo, *CRISPR Seed Resurrection*. Fundação Eugénio de Almeida in Evora, Portugal, 2021. © Ken Rinaldo. Photo: Marta De Menezes

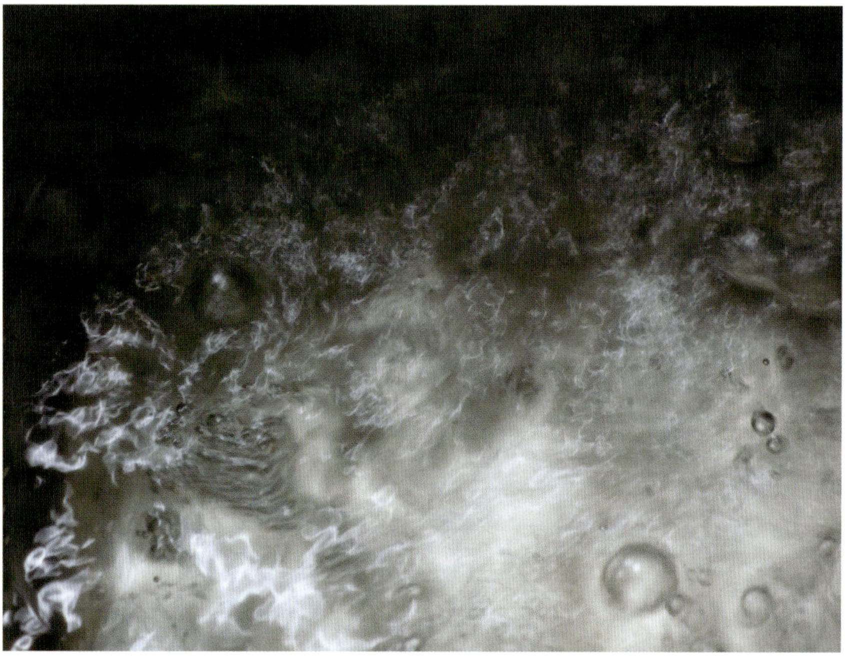

PLATE 30 Meredith Tromble and Dawn Sumner. Detail of boiling water "landscape," *Dream Vortex*, 2011–19. Courtesy of the artists. All rights reserved

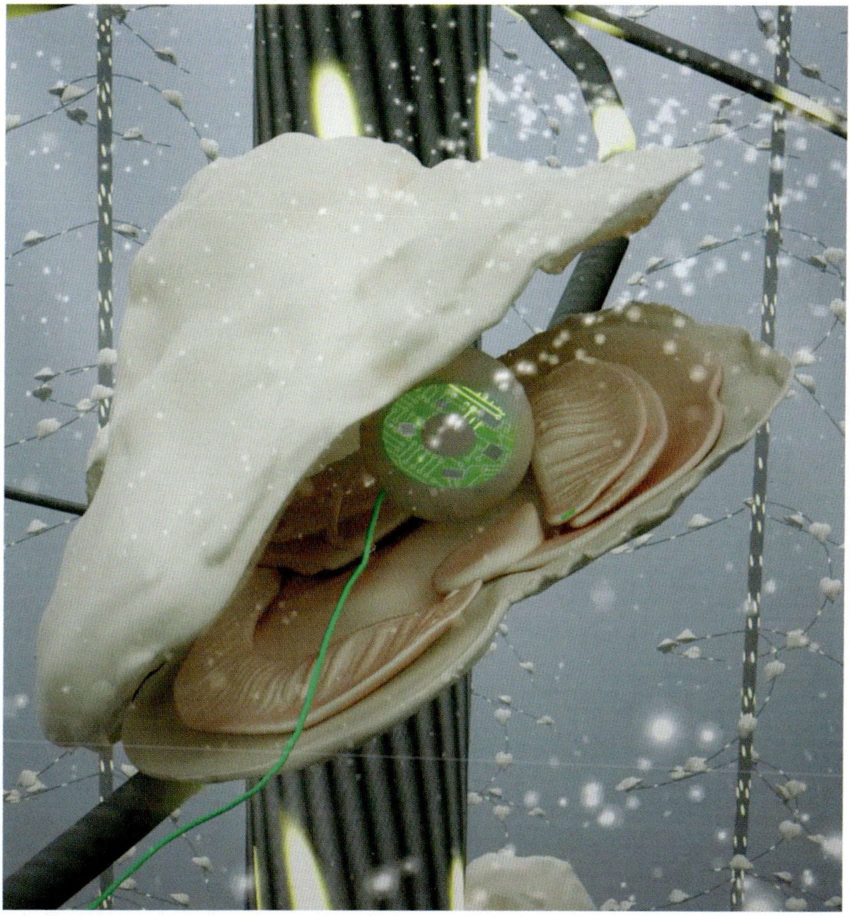

PLATE 31 Image of Aquadisia Networks aquaculture system, 2020. Model by Sputnik Animation. Courtesy of Stephanie Rothenberg

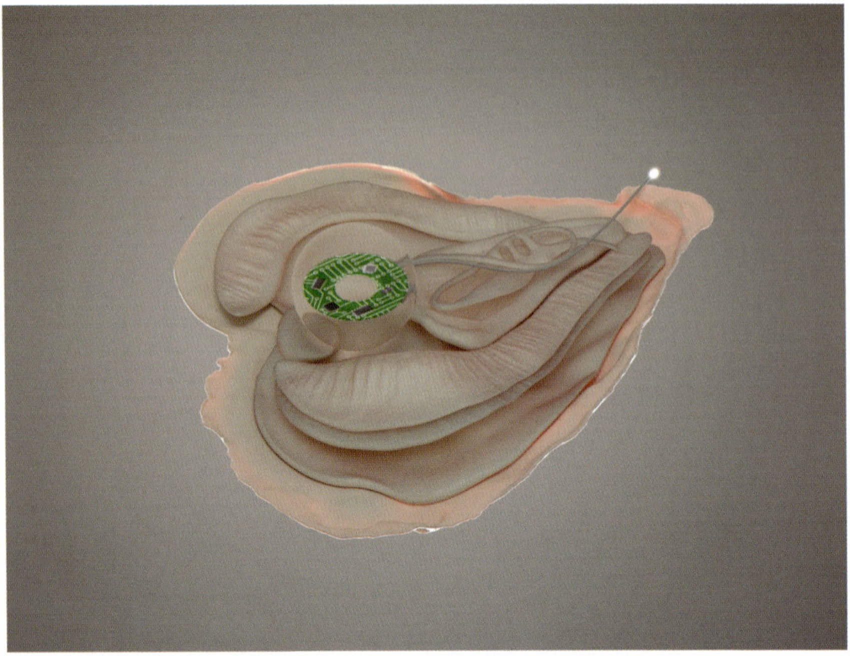

PLATE 32 Image of bioengineered oyster, 2020. Model created by Sputnik Animation. Courtesy of Stephanie Rothenberg

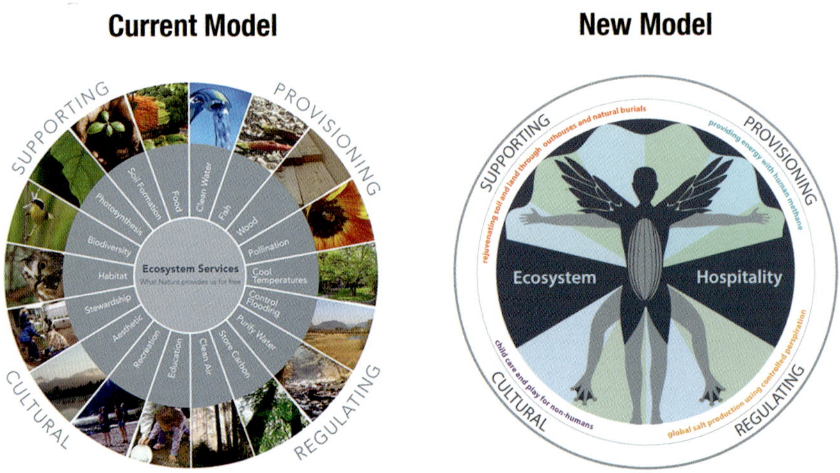

PLATE 33 The new model on the right provides just a few examples of an "ecosystem hospitality." Left image source: IUCN/International Union for Conservation of Nature. Right image courtesy of Stephanie Rothenberg, 2021

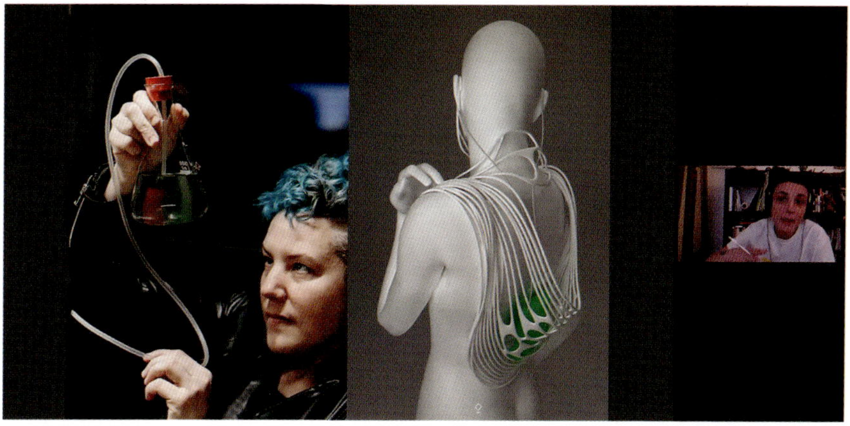

PLATE 34 Fara Peluso, presenting her talk at VIRAL CLOUD, online forum between Art Laboratory Berlin and BioClub Tokyo, 2020, video still. Courtesy of Art Laboratory Berlin

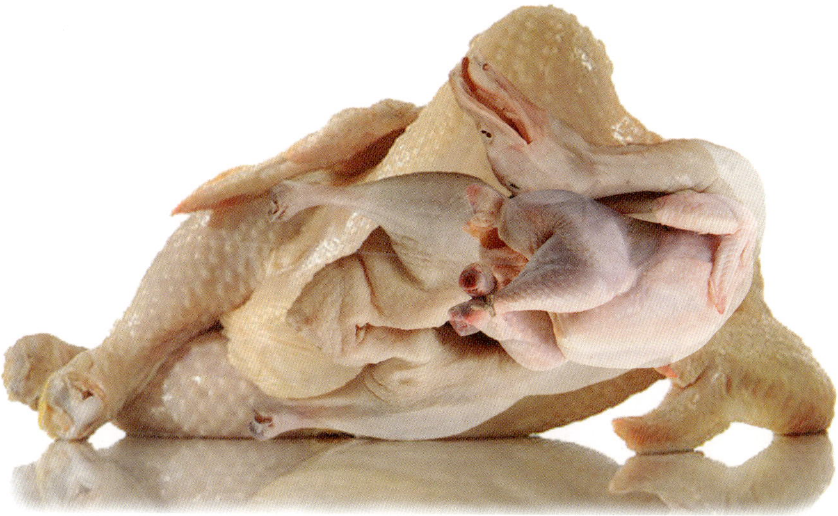

PLATE 35 WhiteFeather Hunter, *Digital montage of found photos of raw turkey, duck, and chicken bodies*, 2021. Courtesy of WhiteFeather Hunter. © WhiteFeather Hunter

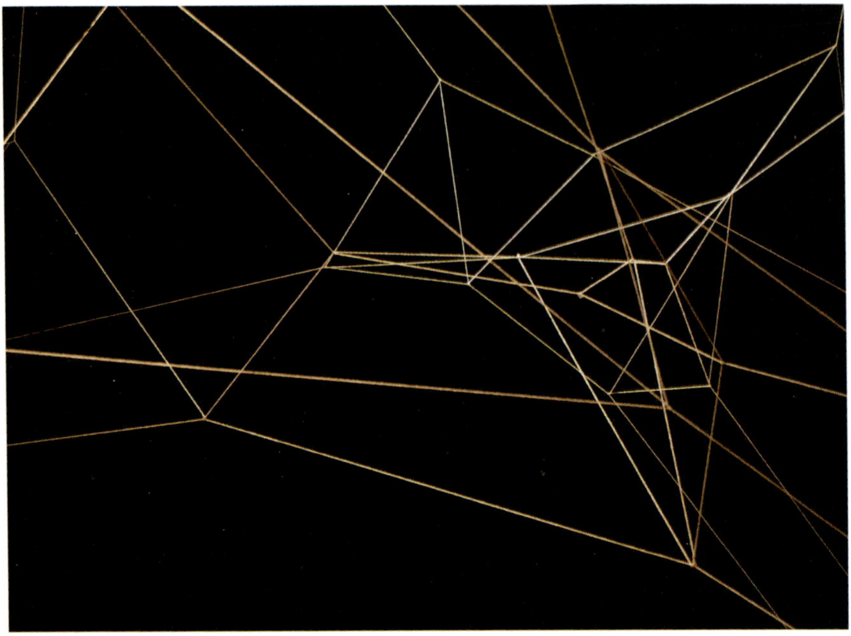

PLATE 36 Marta de Menezes and Kira O'Reilly, *Spider Web*, 2013, installation performance. Courtesy of Cultivamos Cultura, Portugal

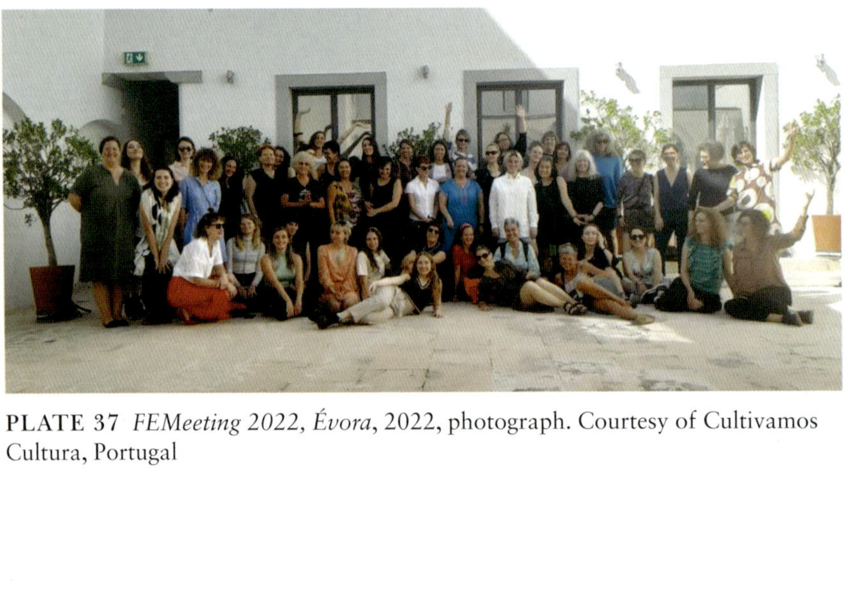

PLATE 37 *FEMeeting 2022, Évora*, 2022, photograph. Courtesy of Cultivamos Cultura, Portugal

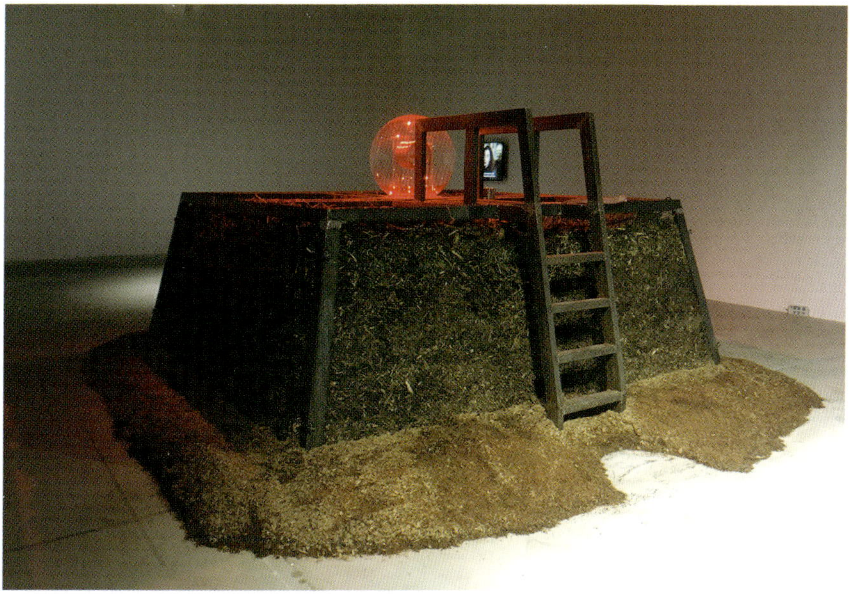

PLATE 38 Oron Catts, Ionat Zurr, and Devon Ward, *Compostubator 0.4*, 2019. From Growing Exhibition, Chronus Arts Centre, Shanghai China. Courtesy of the artists

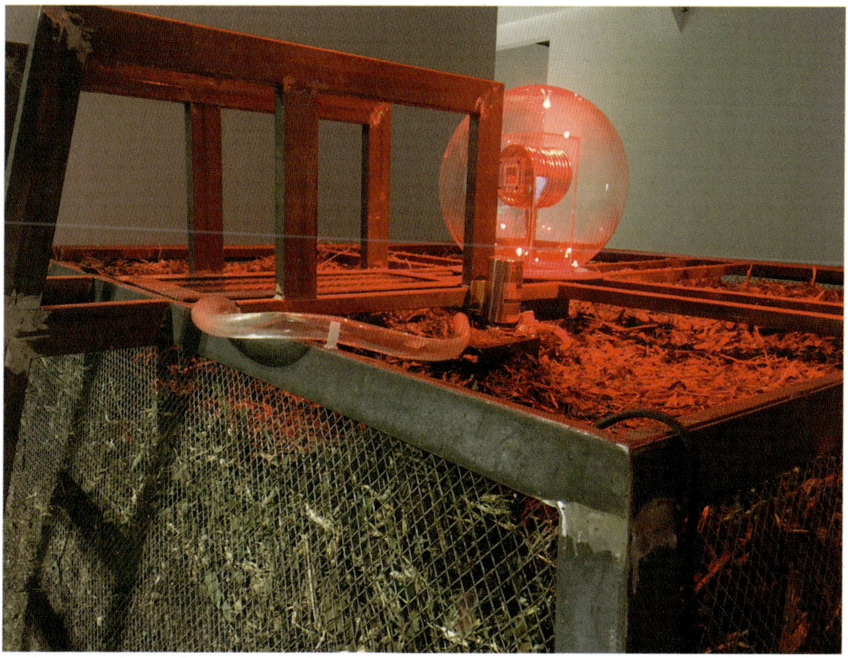

PLATE 39 Oron Catts, Ionat Zurr, and Devon Ward, *Compostubator 0.4*, 2019. From Growing Exhibition, Chronus Arts Centre, Shanghai China. Courtesy of the artists

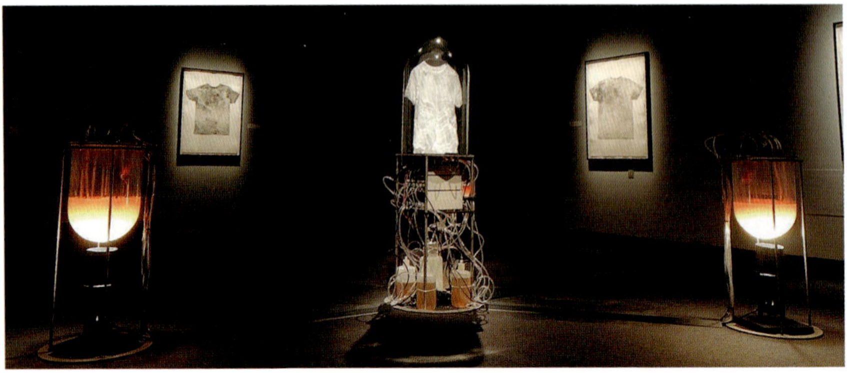

PLATE 40 Paul Vanouse, *Labor*, Burchfield Penny Art Gallery, Multisensory installation, 2019. Photo: Francesca Bond/Buffalo News. Courtesy of the artist

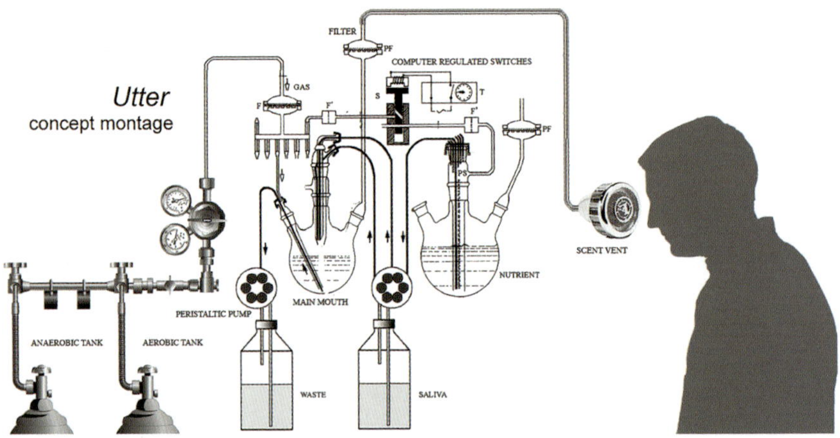

PLATE 41 Paul Vanouse, *Utter* project, 2021, Concept montage: A collage of varied industrial diagrams (based upon an artificial mouth built by G. H. Dibdin and colleagues in 1976) meant to give a general sense of design, not an exact schematic. Courtesy of the artist and Elsevier

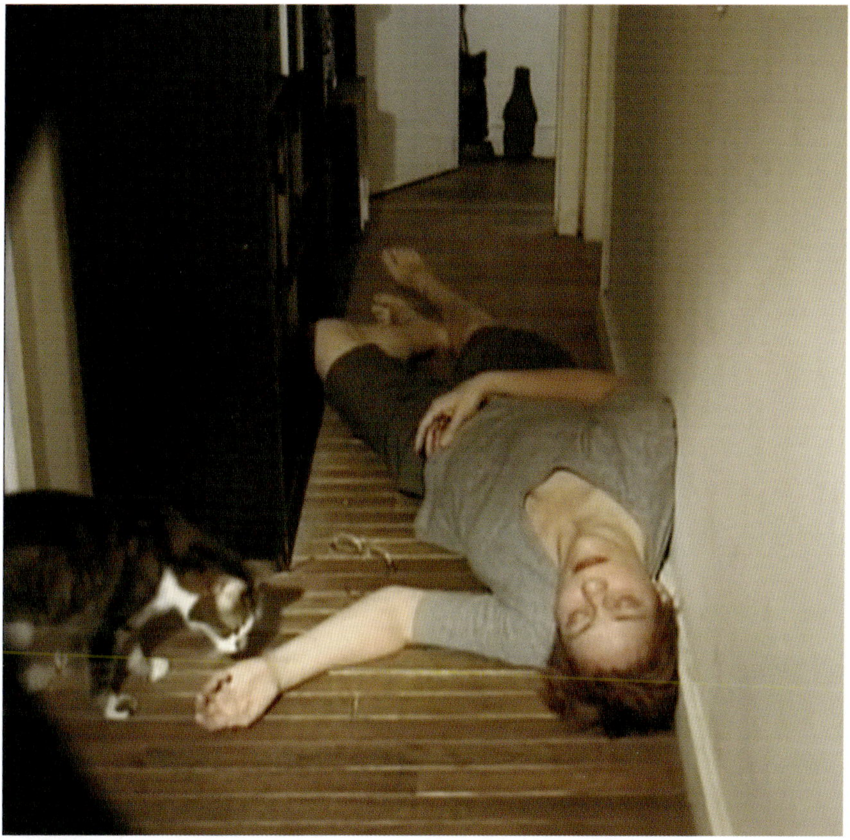

PLATE 42 Kathy High, *Everyday Problems of the Living: Death Poses*, 2000. Courtesy of the artist

Linda M. Montano

DEATH NOTES

In 1988 my mother died. It was a five week hospital death. Colon cancer. I was with her much of that time. These are some of the things that I learned.

1. Ask the dying person questions and wait for verbal/ nonverbal responses i.e. "Should we stop this medicine?" "What do you need?": (I didn't always ask my mother's opinion and now wish I had).

2. Listen, listen, listen.

3. See if they have unfinished business they want to work with, i.e. unexpressed thoughts, feelings, wishes, etc. (My mother said, "I always wished I had written a book," knowing that I would hear her and do it for her. I have.)

4. If they prefer, let them leave consciously, alertly, with awareness. (Because one medicine was stopped, she became comatose. I didn't know that would happen and I missed saying goodbye because of that.)

5. If they agree, play nature tapes in the hospital room. It creates a no panic atmosphere. (I played one and nurses used to come in the room because it felt "good in here").

6. Help the way they need to be helped, not the way you need. (Once I was chanting, praying, laying my hands on her abdomen, teaching her to breathe and my mother, always the comedienne, opened her dying eyes, looked at me with an Imogene Coco-look and said, "Linda, please!")

7. Confess by their bedside. (Clear your heart but do it telepathically. Tell everything you need to tell. Do unfinished business without their hearing. Sotto voce.)

8. If the person is on heavy painkillers, they might change their behavior toward you, positively or negatively. Be ready! (My mother became a hippie on morphine. She pulled me close to her, tried to show me my aura, touched the peach fuzz on my cheeks. We re-bonded.)

9. Get counseling yourself from either friends, 24-hour telephone hotlines, Hospice volunteers, etc. I talked with a Hospice volunteer for 4 years on the phone while my mother was sick and dying. (Hurrah for Hospice!!)

10. Know what "patient's rights" means and use the information appropriately. (I asked one nurse, "If this were your mother would you allow a nurse to take blood and do "vitals" while your mother was comatose and had only a little time left?" The nurse said, "No." I said, "Then please do not take blood and stop all orders at the desk.")

11. Be prepared for each family member and close friend of the dying person to have a completely different response to everything. Emotions are close to the surface; everyone's death anxiety button is being pushed.

12. Use TV, VCR's as teaching tools. (When my mother wanted TV on I turned to cartoons. The flying horse image from My Little Pony did wonders for her attitude and actually kept her pain-free.)

13. Inconspicuously breathe together (match your breathing pattern to theirs), and gently sound the exhalation. (Steven Levine teaches this technique. It's Tibetan and comforting for both patient and caretaker.)

14. Whisper messages near an ear. Keep messages positive and in words they need to hear. (When I said, "Relax, Ma, Mitchell and Karl will take very good care of you," she responded positively. These were friends who were dead and whose company she loved.)

15. Go as far as you can go with the process at the end. (Those last 15 hours when she was comatose, I felt a need to meditate, give her space, not touch her as much. Check out your own situation, it will be different).

16. After the death participate in the process as much as you can given place,circumstances, etc. Help wash the body, close the eyes, etc. (I didn't see the spirit or soul leave. Some people do.)

17. Never, ever judge how you or anyone else mourns or deals with a death. (I demolished and rebuilt an abandoned building for two years after she died. I rarely cried, but my body sweated.)

18. Be watchful for messages, dreams, symbolic visits. (She comes as a butterfly or sometimes a wave of feeling. I continue a dialogue with her.)

19. Daily: Prepare for your death in your own way.

20. Your comments:

Typesetting: Janet Bizzell

PLATE 43 Linda Mary Montano, *Death Notes*, 1988. Courtesy of the artist

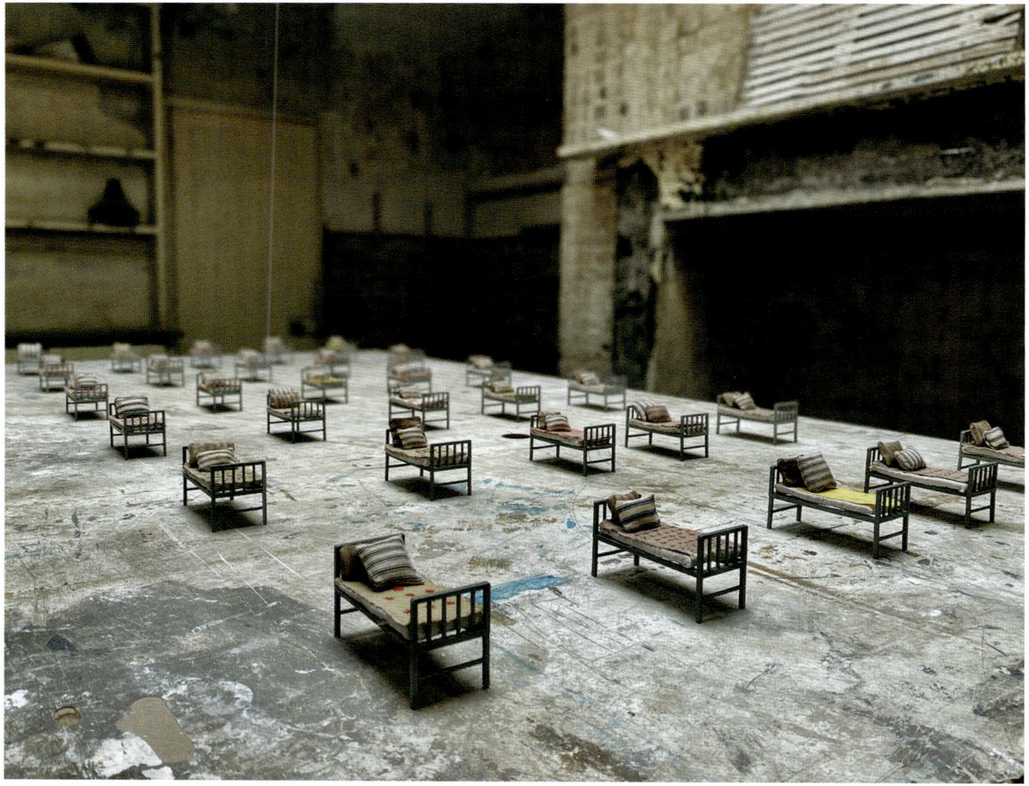

PLATE 44 Anna Dumitriu, *Shielding*, 2020, installation view at the Regency Town House, Brighton, UK. Commissioned by ART/DATA HEALTH. Photograph by Anna Dumitriu. Courtesy of the artist

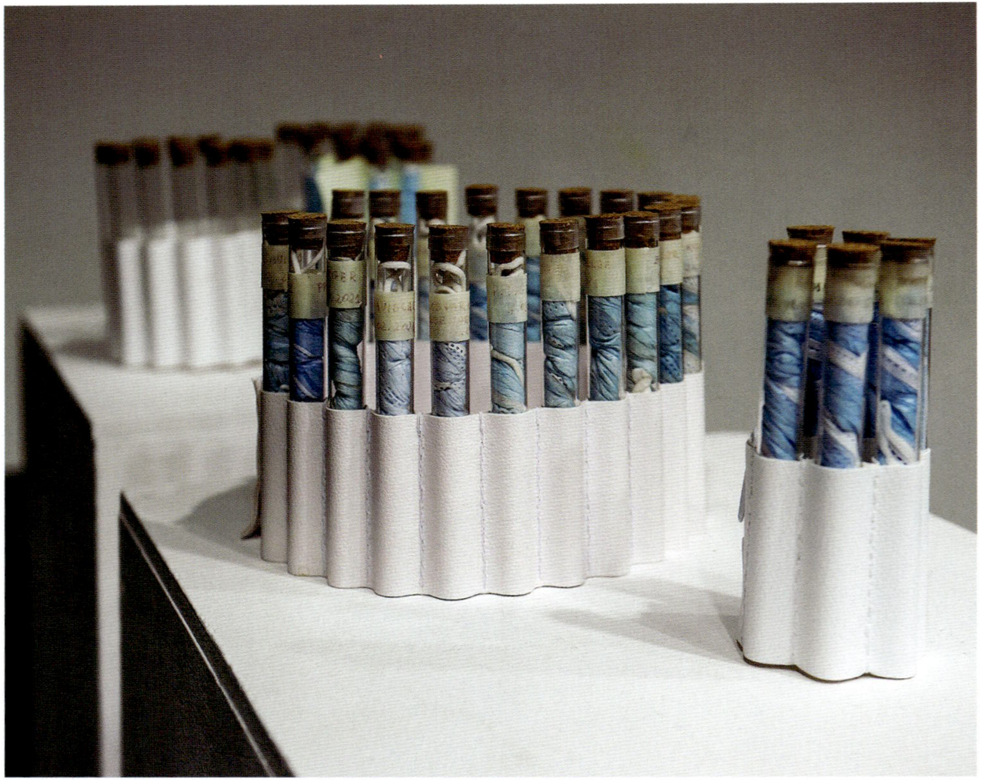

PLATE 45 Karolina Zyniewicz, *Signs of the Times. Collecting Biological Traces and Memories*, documentation of the event at Contemporary Art Centre Łaźnia in Gdańsk. Photo by Adam Bogdan. Courtesy of the artist

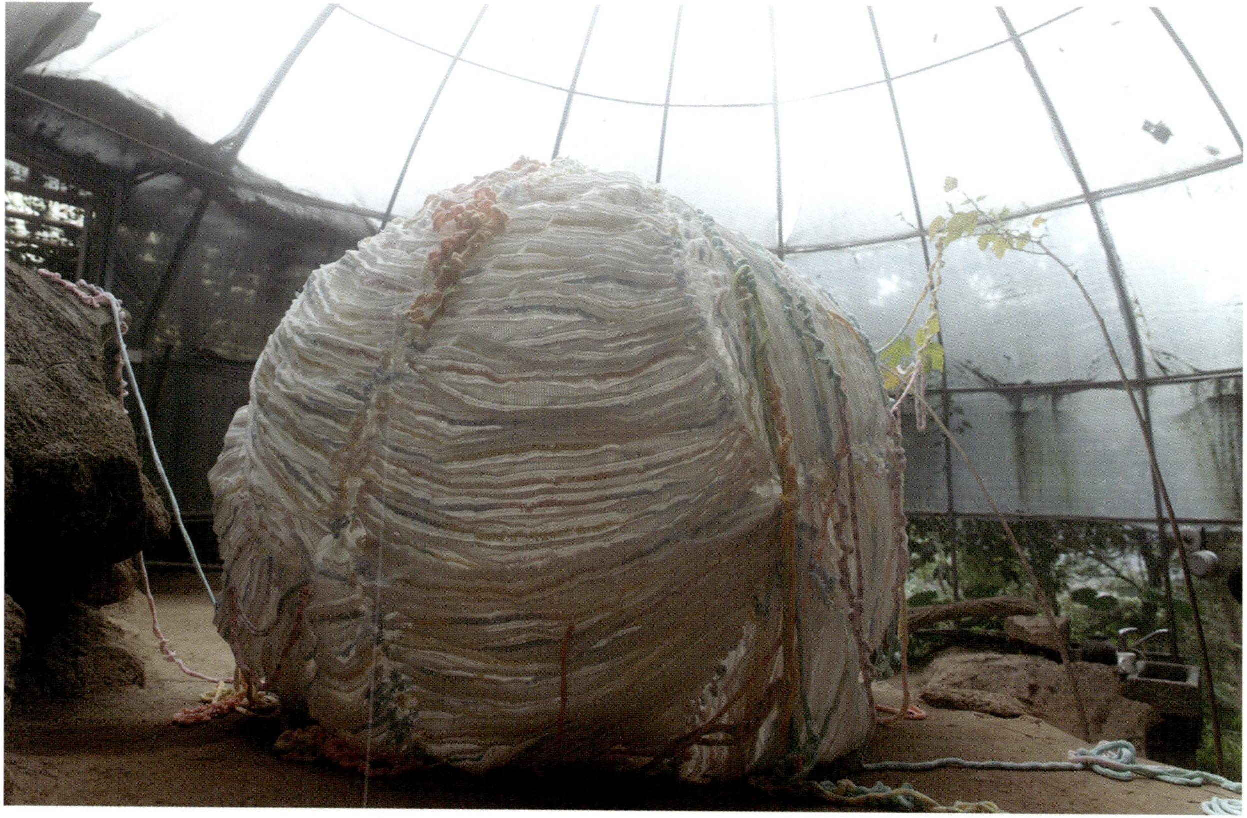

PLATE 46 Pei-Ying Lin, Studies of “Interbeing Trance 1:1,” in the process of becoming part of “the Nature.” Courtesy of the artist

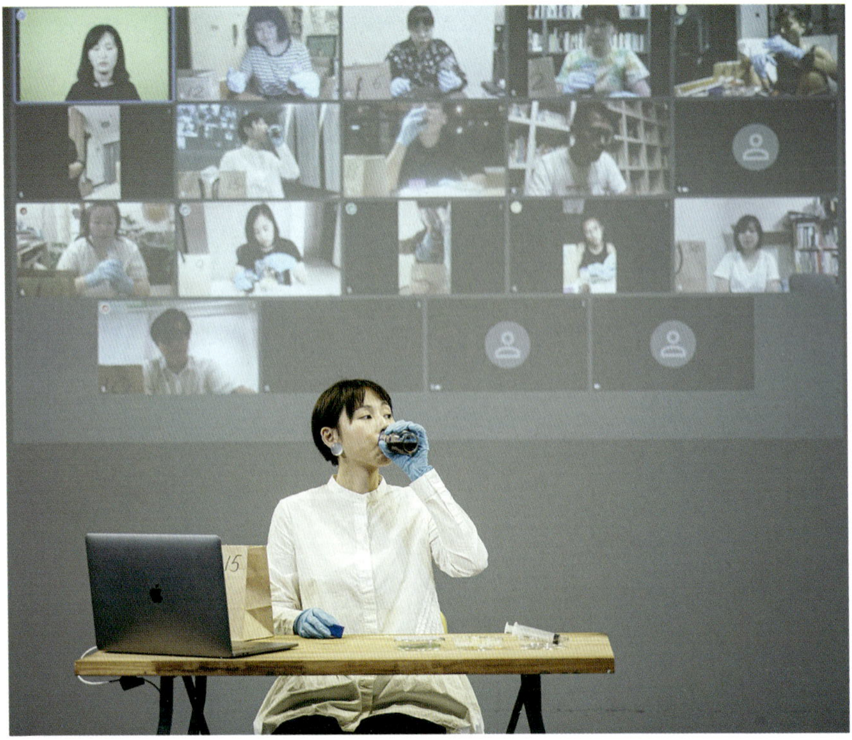

PLATE 47 Pei-Ying Lin, *Virophilia*, Dinner Performance, 2020. Photo by C-Lab Taipei. Courtesy of the artist

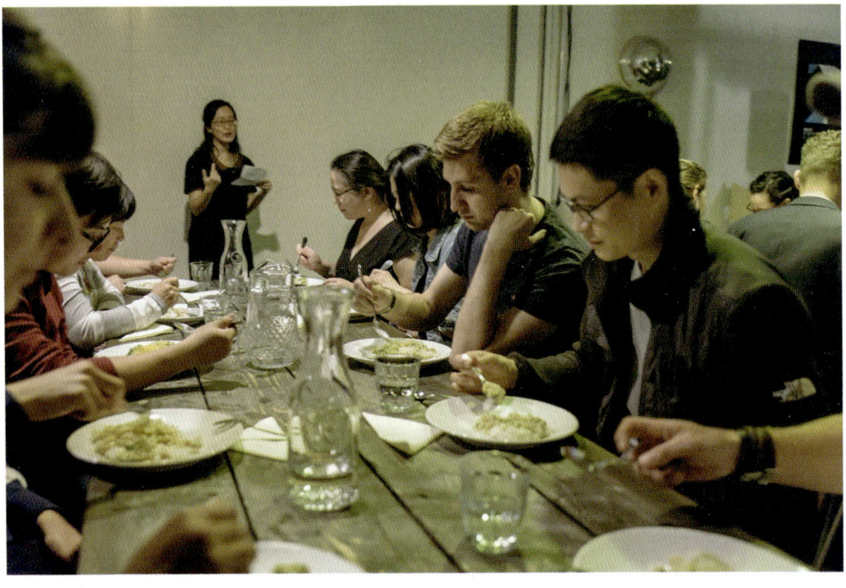

PLATE 48 Pei-Ying Lin, *Virophilia*, Dinner Performance during Cinemasia, 2019. Photo: Armando Ello. Courtesy of the artist

PART FOUR

COVID and the Arts: Practice

16

Thermobiopolitics

Oron Catts and Ionat Zurr

Regulating temperature for optimal biological performance is something that the earth and many of its inhabitants share. A deviation in the temperature range of operation may indicate a problem—be it global heating or expressing one of the symptoms of COVID-19. Sensing, measuring, and regulating temperatures are of ecological, environmental, technological, and biopolitical importance. The application of biological and technological thermoception (the sensing of temperature difference) to control different types of bodies (biological, mechanical, planetary) can be referred to as *thermobiopolitics*. The laws of thermodynamics—whether as epistemology, methodology, or ontology—point to the long-term futility of isolated systems, and how *thermobiopolitics* respect and disrupt bodily boundaries through dynamics of care and control. The first law of thermodynamics, also known as Law of Conservation of Energy, states that energy cannot be created or destroyed in an isolated system. This is true in the use of incubators, which consume biological or electrical energy to produce heat. The second law of thermodynamics states that the entropy of any isolated system always increases. As an interconnected planetary system, life on Earth creates a temporary reprieve from this law, provided an external energy force, such as the sun, is part of the system. Artistic engagement with these issues can generate awareness and new insights into the multifaceted manifestations of thermal regulations.

Epistemology

Most biological organisms possess *thermoception*—the ability to sense changes in heat energy (temperature), both internally and externally, of their bodies. Life evolves through the interplay of the adaptation of traits to changing environmental conditions. Many organisms keep their insides shielded in near optimal conditions for biological function and homeostatic equilibrium to be maintained—these organisms are endotherms. Some argue that life has evolved complex feedback systems on a planetary scale that help to regulate the earth's climate. A disruption of these systems will have a major impact on life on Earth. This is climate change.

The CLAW hypothesis (Charlson, Lovelock, Andreae, et al. 1987) claims that the earth possesses a form of biological thermoregulation: Dimenthylsulphide made from ocean algae is a source of cloud-condensation nuclei, which, if doubled, could counteract global heating caused by the doubling of atmospheric CO2. Lovelock's subsequent Gaia hypothesis (2003) suggests that the earth was living, reconciling perspectives within evolutionary biology and evolutionary geology. Exploring Earth as a whole self-regulating system, Lovelock (2003) writes: "The theory is provisional but provides an intellectual habitat where understanding of the earth can evolve and grow. Perhaps its greatest value lies in its metaphor of a living Earth, which reminds us that we are part of it and that human rights are constrained by the needs of our planetary partners."

One can thus consider the emergence of human-induced global heating as a form of fever. In endothermic organisms such as mammals, fluctuations in body heat are likely to indicate that something is wrong. In the case of a viral infection, the body often responds by raising its temperature.

The change in body temperature is now used as part of the thermobiopolitical theater of control and containment enabled by the thermometer. Thermal imaging cameras and thermometers have become a common part of border crossing in airports and points of entry since the SARS outbreak of 2003. In 2020, they have become a common means to control movements of human bodies in public buildings. In a strange twist of irony, these spaces are generally climate controlled, and as Starosieleski (2014) reminds us, "cooling systems have been mobilized to increase labor productivity; to reinforce existing divisions of race, class, and gender; and to cultivate particular inhabitations of the environment." In our artistic practice, we have presented heating, rather than cooling, systems as places of labor, care, and resistance.

Fever screening using thermal imaging cameras and infrared (IR) thermometers measure body surface temperature and not core body temperature. A growing number of experts point out that measuring body surface temperature tells us little about whether a person actually has fever (Tipton and Mekjavic 2021). Surface body temperature is influenced by many other factors such as weather, exercise, alcohol consumption, etc. Fuller and Mitchell (2020) explain: "Another problem is that skin temperature does not rise during the developing phase of a fever. It falls because warm blood is kept away from the skin. So your skin temperature changes in the opposite direction to your body core temperature." It is interesting to note that many agencies and regulators such as the European Centre for Disease Prevention and Control and the UK's Medicines and Healthcare Products Regulatory Agency warn that surface body temperature is not a reliable form of fever or viral screening method ("Temperature

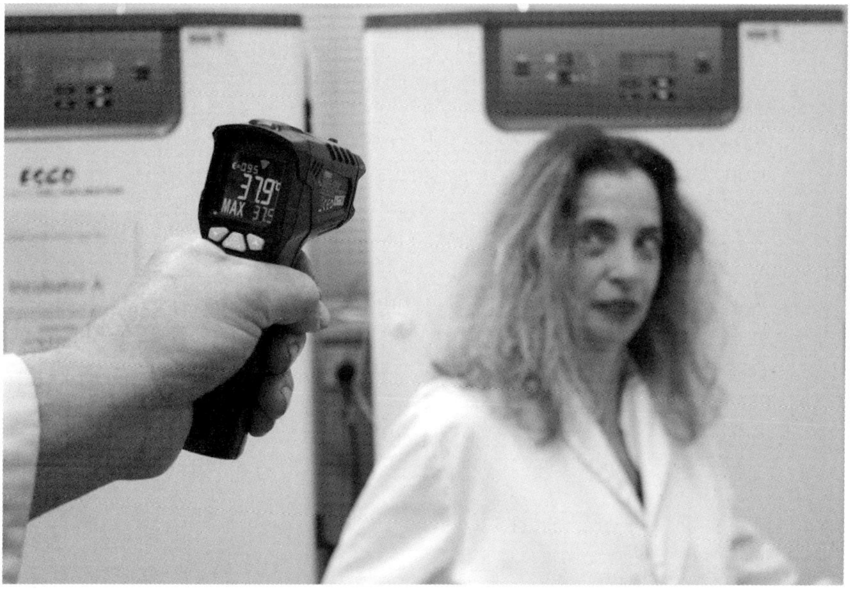

FIGURE 16.1 *Oron Catts and Ionat Zurr,* Thermoception, *2020. From Sensory Orders Exhibition, ŁAŹNIA Centre for Contemporary Art (PL). Courtesy of the artists.*

Checks and Thermal Scanning" 2020).The thermometer must be inserted deep inside a body orifice to give a more accurate estimation of the ever-dynamic thermo-situation. This is an example, both literal and conceptual, of living biological resistance to standardization and technological control. Living biological bodies are dynamic thermobiopolitical resistant agents that require more intrusive technological disciplining.

Methodology

Even though humans can subjectively sense changes in temperature through their own bodily thermoception, it was not until the end of the sixteenth century that an "objective" device was invented. This first technological instrument of temperature sensing was the thermoscope—an air thermometer which measured expansion of air when heated. Also called "weather glass," it is often credited to Galileo in 1593—though, some attribute it to Santorio Santorio (1561–1636), who recorded in 1612 that he developed a device with a numerical scale that indicated relative fluctuations in temperature (Van Helden 1995). In the eighteenth century, physicist Daniel Gabriel Fahrenheit proposed a standardized scale to measure temperature (Balmer 2010); with the freezing point of water set at 32°F, boiling point at 212°F, and human body temperature at around 100°F.

Cornelis Drebbel developed a device that used a feedback system to regulate constant temperature—the thermostat—in 1620 (Tierie 1932). Considered to be the first, first-order cybernetic device, the thermostat is an artificial *thermoregulator* that controls the temperature in specific environment. Its original use was for incubating eggs without hens. Considering humans' ongoing (and accelerating) ambition to control life for their own desires, it is not surprising that the thermostat was used to outsource the care for and control of other life forms to a technological device. Early examples of chicken incubators were often referred to as "artificial mothers." The aim was to hatch chicks without the mother sitting on eggs, enabling hens to continue laying eggs or be used as meat products without interruption. "Through the ability to simulate the hen's brooding functions, it secured the supply chain of chickens and eggs as an industrial commodity" (Leonard and Maury 1879). This could be described as an early example of biological engineering where the feedback-controlled industrialization of animal life has been facilitated by an external, nonbiological technology, as opposed to breeding, grafting, or extracting biological materials from bodies.

The authors of this chapter, who have for three decades explored humans' changing relationships to the concept of life, have been growing cells and tissues in artificial techno-bodies to create semi-living entities. Over the years, their attention extended beyond living cells to the environment which keeps these cells alive. This environment is dependent on thermoregulation. We have developed artworks examining *thermoception* as it relates to incubating (semi)living processes. Humans are not the first organism to use external incubation to control the development of living systems. Here in Western Australia, the Malleefowl has developed a unique thermoregulation system; they incubate their eggs in a large compost mound that they build, and monitor its temperature with their bill and tongue. The male bird will alter the compost's structure to maintain its temperature at 33°C by uncovering the eggs to let air flow when the mound's metabolic activity creates an excess of heat or by adding sand or soil when the temperatures cool down (Seymour 1991).

In a series of artworks, we have developed the *Compostcubator*—a constructed compost pile, which generates heat via microbial metabolic activity that is distributed through a water flow to a tissue culture incubator positioned on top. The heated incubator sustains and supports the growth of animal cells in a tissue culture flask. Rather than using an automated electrical incubator (routinely used in a laboratory) we are employing microorganisms as thermo-labor.

The second experiment, the *Compostcubator* ran at the Science Gallery London's *Spare Parts* exhibition in the midst of the winter when temperatures ran below 0°C. This was a pre-cybernetic incubator, as we did not use a thermostat to regulate the temperature in the incubator. We relied on the fact that the core of the compost pile tends to maintain a steady temperature,

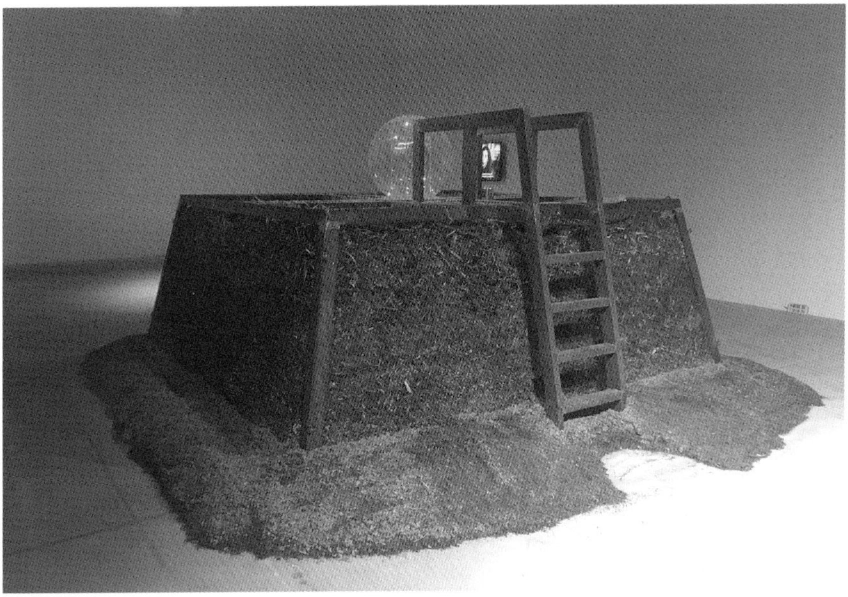

FIGURE 16.2 *Oron Catts, Ionat Zurr, and Devon Ward,* Compostubator 0.4, *2019. From Growing Exhibition, Chronus Arts Centre, Shanghai China. Courtesy of the artists.*

of around 60°C, as the microbial community self-regulates its heat producing activity. We set the rate of flow of the water that ran through the core of the compost to heat and maintain a steady and comfortable environment for the cells in culture. Our *Compostcubator* "body" provided life support to scar tissue cells (mouse connective tissue) at the required 37°C. The shape of the installation drew on the architectural features of ancient places of worship, early childbirth training devices (obstetric phantom), and on the nesting habit of the Malleefowl bird. Together with the Faculty of Life Sciences and Medicine at King's College London, we demonstrated that disembodied cells can be sustained alive through thermo-regulation of microbial metabolic activity.

We have presented heating systems as complex relations of labor, care, resistance, accidents, life, death, and decay. An automated electrical incubator simulates a body in an isolated system. It acts as a homeostatic surrogate for life out of context and out of agency. *Compostcubator* complicates this apparatus by making it into a living system—a messy, permeable "biological device" that is an assemblage of different species. The work plays with and critiques notions of control, optimization, interdependence, and multispecies ecologies. The *Compostcubator* also acts as a metaphor for the biologically regulated global climate. When presented for the first time in Perth City Farm, Australia, the cells were cooked rather than incubated, due to the strong, unfiltered solar radiation.

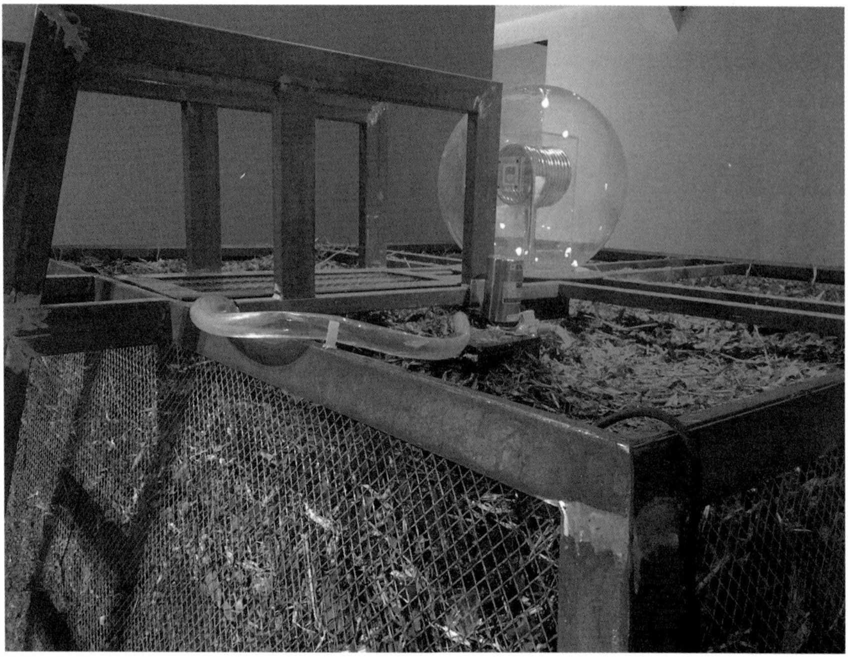

FIGURE 16.3 *Oron Catts, Ionat Zurr, and Devon Ward,* Compostubator 0.4, *2019. From Growing Exhibition, Chronus Arts Centre, Shanghai China. Courtesy of the artists.*

Extreme thermo-fluctuations can be disastrous and lead to death. In the case of global temperature rises, "At plus-4 degrees, *hordes* of climate refugees would flee famine and extreme water scarcity. At plus-5 degrees, climate refugees would number in the tens of millions as massive uninhabitable zones spread" (Ostrow 2008). Global heating and personal warming become entangled in a strange theater of failures of temperature regulation. However, while accurate measurements of planetary temperatures are accepted as an indication of global heating, action seems to be slow and ineffective. Contestable and scientifically dubious measurements of human body temperatures are becoming an effective biopolitical weapon of control, over movement of human bodies across borders and boundaries.

Ontology

Contemporary biology is increasingly becoming a discipline of isolation and control. Since the outset of the COVID-19 pandemic, the rhetoric of confinement and domination has extended to the management of the spread of the virus in human society—social distancing, population lockdowns, and so on.

In the lab, isolating life from context for safety or efficiency reasons seems to bias much of the rhetoric and practice in this field to focus on controlling living systems at the molecular level. The same applies to COVID, caused by the less than 30,000 RNA bases of the SARS-CoV-2 virus. Knowing the RNA sequence of the virus does not tell us much about the economic and social turmoil it has inflicted on humanity. SARS-CoV-2 can only be understood in the broader context of its host. Life which is removed out of its original context has to integrate into a newfound biological (i.e., a SARS-CoV-2 virus in a human host cell) or artificial (i.e., cells in an incubator) "body" which simulates and regulates the internal conditions of the original body from which it was removed. Here we give homage to the underrepresented hosting body rather than the DNA/RNA information. We are using the thermostat and the thermometer as sensory objects with a changing sensory order.

While the thermostat regulates temperatures, the thermometer states them. The thermometer has today become an evocative, contestable, and political inscription device. It provides data of our ever-heating planet, validating or disproving climate change models and skeptics, and symbolically allows us to see whether human bodies became incubators for hosting organisms out of context—the coronavirus. Our current thermally disrupted times may become a fertile milieu to "incubate" different metaphors about what life is: metaphors that give center stage to thermodynamic, sensorial, contextual hosting bodies—from viruses, cells, whole organisms, to the earth—the holobionts.[1]

Living biological systems are agents of fluctuation and interdependence, which cannot thrive in long-term isolation without some warmth when it gets too cold or without cooling mechanisms when heat becomes unbearable, but different bodies have adopted different thermoregulation systems. Self-regulation of a body is relative to the bodies within it and outside of it. Measuring heat energy is becoming a tool for politics and control. *Thermobiopolitics* needs to be explored, unpacked, and used to question bodies from the viral to the planetary scale.

Note

1 The assemblage of the host and microbial community.

References

Balmer, R. T. (2010). *Modern Engineering Thermodynamics*. Academic Press, 9. https://www.arma.org.au/wp-content/uploads/2017/03/Modern-Engineering Thermodynamics.pdf

Charlson, R., Lovelock, J., Andreae, M., et al. (1987). "Oceanic phytoplankton, atmospheric sulphur, cloud albedo and climate." *Nature* 326: 655–61.

Fuller, A., and Mitchell, D. (2020). "So you think investing in fever screening can curb the spread of COVID-19? Think again." *The Conversation*. https://theconversation.com/au

Leonard, S., and Maury, J. (1879). "Improvement in incubators." Patent No. 217,121, July 1, 1879.

Lovelock, J. (2003). "Gaia: The living Earth." *Nature* 426: 769–70.

Middleton, W. E. K. (2003). "History of the Thermometer and Its Use in Meteorology." Johns Hopkins University Press.

Ostrow, J. (2008). "Six Degrees' Charts Climate Apocalypse in HD Television." *Denverpost.com*, February 8. http://www.denverpost.com/ostrow/ci_8190284

de Réaumur, R.-A. F. (1750). "The Art of Hatching and Bringing up Domestic Fowls; by Means of Artificial Heat."

Seymour, R. S. (1991). "The Brush Turkey." Scientific American 265(6): 108–15. http://www.jstor.org/stable/24938837

Starosieleski, N. (2014). "Calibrating a Sense of Environment: Artificial Cooling and the Biopolitics of Thermoception." Anthropocene Feminism Conference, Center for 21st Century Studies, University of Wisconsin-Milwaukee. https://c21uwm.com/anthropocene/abstracts/

"Temperature Checks and Thermal Scanning" (2020). Public Health England, June. https://www.england.nhs.uk/south/wp-content/uploads/sites/6/2020/09/2B-Info_sheet_Temperature_COVID19_v1.0.pdf

Tierie, G. (1932). *Cornelius Drebbel*. H. J. Paris.

Tipton, M., and Mekjavic, I. (2021). "Temperature scanners aren't good at telling who has COVID-19—here's how to fix that." *The Conversation*. https://theconversation.com/au

Van Helden, A. (1995). *Galileo Project*. Rice University. www.galileo.rice.edu.

17

Utter

On the Matter of Human Emissions

Paul Vanouse

Foul words are but foul wind, and foul wind is but foul breath, and foul breath is noisome; therefore I will depart unkissed . . .

WILLIAM SHAKESPEARE, *MUCH ADO ABOUT NOTHING*, ACT V SCENE 2

The terms "utter" and "utterance" are generally understood in a verbal sense, but they stem from the more physical Old English root, "utter"—to put forth or send out (OED 2022). When we utter, we release a moist, odiferous cloud of saliva, molecules, microorganisms, and viruses. This essay seeks to integrate literary and philosophical models of utterance, with this less explored, material facet, foregrounded and implicated by the COVID-19 global pandemic, which challenges our existing social and philosophical paradigms, including how we privilege language over matter.

These concepts are embodied in an in-progress, multisensory artwork, which aims to place historically marginal, hidden, and abject qualities of human breath back into the realm of aesthetic contemplation. Utter comprises a series of functional "artificial mouths," biotechnological artifacts that will produce the damp and fragrant clouds of human breath. The networked apparatuses incubate a myriad of bacterial species native to the human oral cavity in-vitro and release the scented clouds through a series of vents along the apparatuses' perimeter. The essay and artwork are intended to cross-inform this investigation into the material, nonhuman, and nonauditory aspects of speech.

Beyond Semiotics

> The living utterance, having taken meaning and shape at a particular historical moment in a socially specific environment, cannot fail to brush up against thousands of living dialogic threads, woven by socio-ideological consciousness around the given object of an utterance. (Bakhtin 1981: 276)

Mikhail Bakhtin expands and liberates the concept of utterance, from its usage in linguistics and the philosophy of language. He is critical of linguists such as Ferdinand de Saussure, who viewed language as a formal, closed system that can be analyzed apart from the messy dialectics of everyday speech. Bakhtin viewed such accounts as fundamentally hegemonic, serving as a centripetal force, suppressing the heteroglossia, or varieties within a linguistic code, of everyday conversations (Robinson 2011). Conversations contain pauses, emphasis, intonation, gestures, and often multiple dialects, which are understood differently depending upon their historical context and who is speaking and listening. He explains that for words, propositions, and grammar rules to become complete utterances and linguistic acts, there needs to be a "supplementary element" that remains "inaccessible to all categorisations and linguistic determinations, one that linguistics cannot grasp" (Lazzarato 2009: 1). Maurizio Lazzarato writes that for Bakhtin, grammatical structures and statements are merely "technical signs" with potential meaning. The transformation into complete utterances is achieved through "pre-individual affective forces and social and ethico-political forces that whilst being external to language are actually inside the utterance" (1). The utterance is not simply a language act from one subject to another, but an intentional and ethical heteroglossial act containing a polyphony of languages and dialects, modulated by inflection and intonation, from one culturally situated subject awash in a sea of names, toward another subject similarly encoded and embedded, who may make meaning from it (Bahktin 1981).

In the spirit of Bakhtin, my intention with this project is to expand the scope of how we consider the utterance, particularly the material aspects and the process of uttering. Since Bakhtin's time, the critique of early-twentieth-century linguistics and the more recent semiotic turn in the humanities has emanated from many fields including science and technology studies, feminist studies, and postcolonial studies. Actor-network theory scholars, such as Bruno Latour, use the term "semio-material" to flatten human/nonhuman oppositions and hierarchies in their social models (1999). Rather than in binary opposition, contemporary science, technology and society practitioners describe "naturecultures" (Haraway 2015). Karen Barad uses the term "material-discursive" to undo the binary opposition between representations and objects. Barad connects both through their mutual temporalities, unfixedness, and continual state of becoming (2003).

My usage of terms such as "semio-material" and "material-discursive" are intentionally inexplicit because in a project such as *Utter*, ideation and materialization are co-dependent and mutually-informing. The more precise "natureculture" of the semio-material cloud that is emitted from the *Utter* apparatus is still in development. However, the concepts are provisionally useful to avoid recapitulating prejudicial hierarchies of cognitivism regarding the material dimension of human speech, such as when we distinguish the scent of an utterance as a lower, base, animalian remnant that clings to abstract higher purposes of speech.

Emissions

The verb "utter" has roots in Middle German, *ütern*, *üteren*; Middle Dutch *uteren*; Anglo-Norman, *uttrer*; and Anglo-Latin, *utterare*, and has been documented in English literature since the fifteenth century. Its English usage has encompassed both speech and material action—for example, "to issue or send forth," "to publish," "to put or thrust forth, shoot or urge out; to discharge, emit, eject, exhale," and "to disclose or reveal (something unknown, secret, or hidden)" (OED: utter, v: 2023). The adjective "utter" in contemporary usage can refer to "complete" or "absolute", as in "utter darkness," or "outer or exterior part; manifesting on the exterior" (OED: utter, adj.: 2023). This abbreviated etymology suggests an expansive conception of the utterance that encompasses speech and material emittance, manifestations as well as declarations and distinguishes it from that which remains inside.

When we utter, we emit an odorous mist containing molecular compounds and microbes. Coughing and more forceful speaking also produce larger droplets with potentially more endogenous materials that are more limited in range. Molecular compounds include a complex mixture of volatile and nonvolatile molecules that are influenced by our genetics, health, diet, and stress (Whittle et al. 2007). Each of these molecules is both semio-material and multidimensional, signifying the presence of certain microbial configurations and narrating varied biochemical processes taking place in the oral, pulmonary, and gut cavities. Many body odors have evolved to be olfactory messengers, which convey information between individuals, forming narratives that may be used to diagnose illness, to evaluate food and environmental toxins, and recognize kin and potential predators (2007). For instance, the aroma of acetone or other ketones in breath may stem from acids produced in the liver and signal a diabetic condition, while it might also stem from a high-fat "keto" diet, fasting or excessive alcohol intake (Danzinger 2020). The odors emanating from our mouths are multifaceted. They may arise from chemicals in the blood released when they pass along the blood-air barrier in the lungs, from organs and processes within our gut, our teeth, gums or the upper surface of the tongue, a site of bacterial plaque development that is a principal source of halitosis-related volatiles (Whittle et al. 2007).

In the early days of the COVID-19 pandemic, virologist Julian Tang warned us: "If you can smell what someone had for lunch—garlic, curry, etc.—you are inhaling what they are breathing out, including any virus in their breath" (Murphy 2020). Exhaled air is an aerosol containing endogenously generated droplets (Bake et al. 2019). Studies have probed the microbial content of human breath aerosols by collecting Exhaled Breath Condensate and analyzing with whole community DNA sequencing (2019). The droplets in these aerosols have a maximum diameter of 4 um and an average particle size of less than 1um. Bacterial cells in exhaled air are limited to tiny varieties, around 1 um, whereas most viruses are very small, 0.02 nm to 0.5 um and studies confirm influenza variants (Xu 2012), H1N1, and COVID-19 can be found in human breath. Thus, our utterances contain not merely nonliving, chemical materials, but also biochemical agents that also exist in constantly evolving relational states within and without the human body.

Entanglements

> The relational capacity of the posthuman subject is not confined within our species, but includes all non-anthropomorphic elements. Living matter—including the flesh—is intelligent and self-organising, but it is so precisely because it is not disconnected from the rest of organic life. (Braidotti 2013: 6)

When Rosi Braidoti describes the posthuman subject, she articulates a challenge to the long-standing humanist anthropocentric premise that the human body is an autonomous entity. The post-anthropocentric body is host to a myriad of flows, forces, metabolisms, and behaviors. It is the microbes living upon or within "our" human tissues and fluids, "our" microbiota, that orchestrate many of these activities.[1] In parallel with the development of whole-community DNA sequencing technologies, the late 1990s witnessed an emerging awareness of the overwhelming diversity within the human microbiome—the trillions of bacteria, archaea, fungi, and other eukaryotes living on and inside us, forming complex ecosystems occupying the many ecological niches provided by human bodies. This revelation was quickly followed by an explosion of research into the human microbiome, revealing complex microbial diversity within these communities and often complex networks of interactions between community members and their human hosts (Marchesi 2014).

My 2019 project, *Labor* (Figure 17.1), was an exploration of the microbial basis of sweat, typically considered almost uniquely human. In the artwork, however, there are no people involved in making the smell—it is created by bacteria propagating in three bioreactors in the exhibition space. Each bioreactor incubates a different species of human skin bacteria

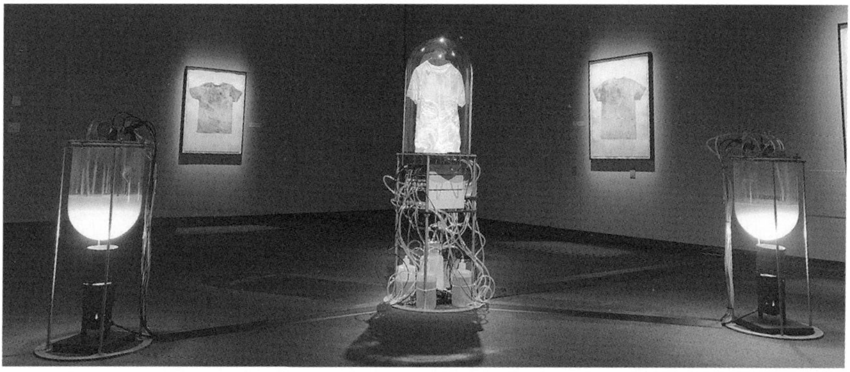

FIGURE 17.1 *Paul Vanouse,* Labor, *2019. Burchfield Penny Art Gallery, multisensory installation. Photo: Francesca Bond/Buffalo News. Courtesy of the artist.*

responsible for the primary scent of sweating bodies: *Staphylococcus epidermidis*, *Corynebacterium xerosis*, and *Propionibacterium avidum*. In each exhibition, these disembodied scents are channeled into a bell jar before diffusing into the room. Human sweat in itself is odorless: it is these bacteria feeding upon the components of sweat that creates volatile, odiferous chemical compounds that we associate with it (Vanouse 2019). In *Labor*, humanness was a phenomenon that was not fundamentally the property of a human subject, but rather of a more complex intra-action between significations of humanness and microbes associated with the human body.

Artist-Scientist duo Lucy Strecker and Klaus Spiess have probed symbiotic interactions within the human body, suggesting that the microbial dimension of one's oral cavity relates directly to phonemes and coevolves with languages. During their *Microbial Keywording* performative artwork, they harvest oral yeast microbes from audience members, which they maintain in-vitro. Phonemes—the smallest unit of meaningful sound—spoken by speakers at the event drive pumps that add corresponding pheromones to the yeast growing medium. The artwork builds on post-humanistic performance theories that question matter-language dichotomies and scientific studies on microbial cellular memory's role in anticipating future events (Spiess and Strecker 2020: 56). In symbiotic linkage between materiality and signification, microbial diversity within the mouth promotes linguistic memory and vice versa.

Their work, linking the performance of speaking with living matter, draws upon Barad's theory of Agential realism. Barad's proposal presents an alternative ontological framework that dismisses a distinction between "words" and "things," replacing this opposition with a more fundamental unit: the phenomena: "dynamic topological reconfigurings / entanglements / relationalities / (re)articulations. And the primary semantic units are not "'words'" but material-discursive practices through which boundaries are constituted" (2003: 818). By centering upon the phenomena as that which breaks this static dualism between observer and observed, she also refutes the idea of causality with that of relationality (815).

In this ontology, rather than a traditional "chicken and egg" problem of causation, phonemes and the microbes in the human oral cavity (and also in the work of Strecker and Spiess) exist as inseparable agentially intra-acting components, neither being the cause nor the effect of the other. Words and things, twin relata, within the speaker's mouth are in a dynamic relation, preexisted by the phenomena itself.

The *Utter* Project

The *Utter* (Figure 17.2) project stems from themes of the *Labor* project, including the production of the qualities of humanness (human smell) by nonhuman actors living on and within the human body and the possibility of this humanness being produced in-vitro. It shares with Strecker and Spiess an oral-fixation and the conception of an underexplored relationship between matter and signification, building upon Barad's theory of Agential realism. *Utter*, however, focuses not only on the oral cavity, but what emanates from it, how it is perceived, and (after Bakhtin) how its meaning is multidimensional, relational, and historical. For instance, during the height of the COVID pandemic the scent of humanness has been interpreted by some as the scent of danger, whereas ethanol and the synthetic perfumes

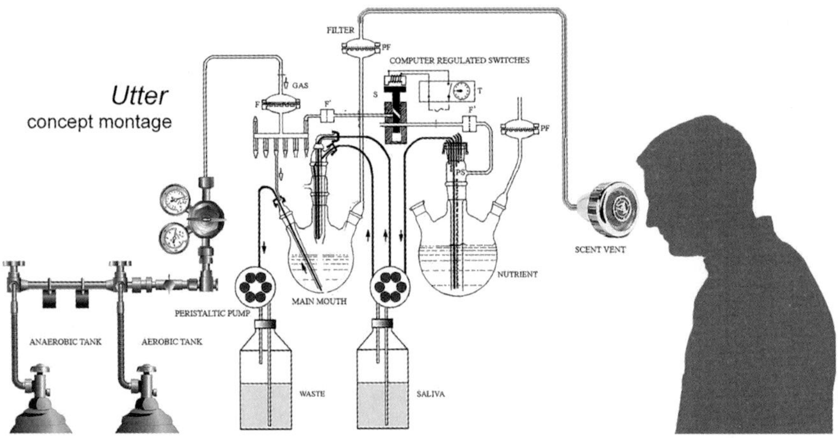

FIGURE 17.2 *Paul Vanouse,* Utter *project, 2021, concept montage: A collage of varied industrial diagrams (based upon an artificial mouth built by G. H. Dibdin and colleagues in 1976) meant to give a general sense of design, not an exact schematic. Courtesy of the artist and Elsevier.*

mixed with it (as found in the now commonplace hand sanitizers) were scents of safety.

Utter is a multisensory installation artwork in-progress, designed alongside scientific collaborator, oral biologist Stefan Ruhl. *Utter* is planned to comprise a series of apparatuses that produce moist, odiferous emissions, reminiscent of human breath. The physical forms of the apparatuses will arise from their function to produce a multitude of distinct, warm breath odors that may be individually experienced by audience members. Each apparatus may stand alone, or link together in a network. Each apparatus is planned to contain designed vents from which odors emanate for viewers to individually inhale and assorted glassware, pumps, valves, filters, tubes, electrical controllers, and other hardware. It will be filled with nutrients, microbes, and artificial saliva. The installation will audibly pulsate and exhaust with the system's mechanical functions.

Artifice

Utter "performances" are planned to be orchestrated via a material network of "artificial mouths," designed at the Coalesce Center for Biological Art at the University at Buffalo. The artificial mouth is an in-vitro technology that has progressed from simple incubation apparatus at the end of the nineteenth century to highly sophisticated, computer-controlled models today (Tang et al. 2003) (Figure 17.3). The complexity of the oral environment as well as the ethics of human studies has led to the need for these artificial models to better study oral diseases in humans (2003). Another antecedent to the mouths planned for *Utter* is called the "Mouth to Nose Merging System," used experimentally in 2012 to isolate the scent of chewed food from other sensations, such as texture and taste (Charles et al. 2013). This system involves using "*in vivo* olfactory receptors of a judge's nose"; however it doesn't focus on microbial scents, but fresh foods, such as apples (2013). The *Utter* project entails designing a custom artificial mouth system, informed by historical and contemporary models, yet optimized for microbial scent production and olfactory perception. The system is conceived to contain multiple vented openings and each of these openings are planned to produce not an individual, nor ideal, nor normative human oral scent, but rather a promiscuously comingled emission.

Barad designates the apparatus not merely as an object to facilitate the measurement of phenomena, but as phenomenon itself (2003). Apparatuses are not static but are "dynamic (re)configuring of the world" (2003: 816). Hans-Jörg Rheinberger has described the experimental set-up as temporary, historically situated, and tied to a process: "What distinguishes experimental systems and epistemic objects from them is a systemic configuration that in its very set-up is not only precarious, but is designed to be so, and a kind of object that in its very constitution carry with it a certain measure of

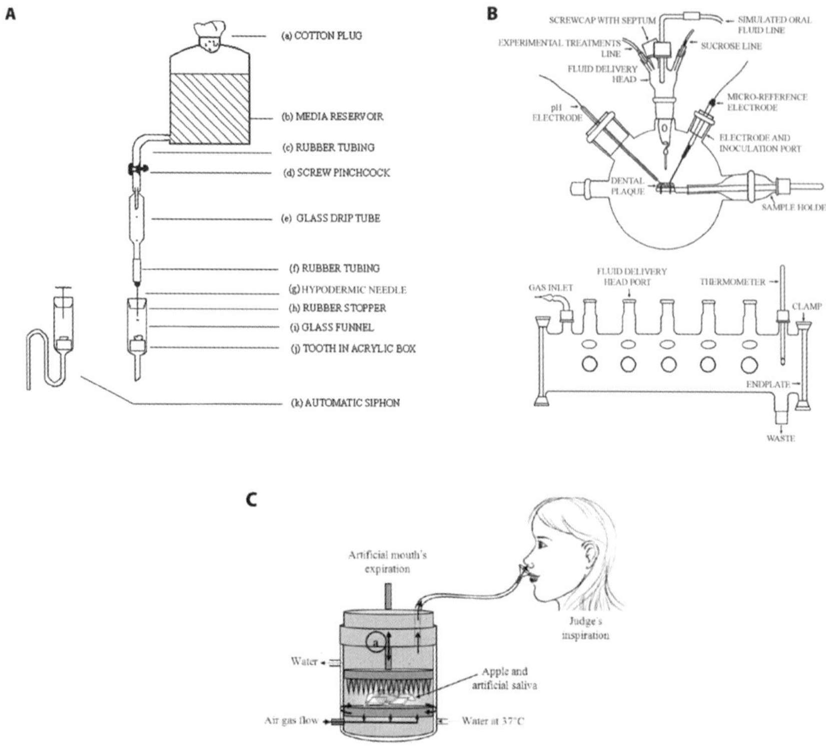

FIGURE 17.3 *(a) Schematic diagram of the artificial mouth designed by Pigman et al. 1952. (b) Schematic diagrams of the artificial mouth designed by Sissons et al. 2000. The lower panel represents a longitudinal section of the culture chamber. (c) Schematic diagram of the "Mouth to Nose Merging System" developed by a joint research team from L'UNAM Université and Université de Caen France. (a) and (b) Tang et al. 2003; (c) Charles et al. 2013. Courtesy of Elsevier.*

under-determination" (2019: 238) Likewise, the *Utter* apparatus is not conceived to be fixed, nor objective, nor timeless, but rather designed to actively coproduce phenomena, intended for broad interpretation.

Outer and Other Space

According to art historian Caroline Jones, "Smell, at least since Locke, Kant and Condillac, has been relegated to philosophical abjection, with fragrance, odor, scent, aroma, perfume and stench all placed at the bottom of the epistemological hierarchy" (Jones and Arning 2006: 12). Visuality, by contrast, has enjoyed privileged status, even as its objectivity has been challenged (Jay 1993). However, there have been several influential scent works produced in the past two decades, particularly those of Norwegian artist Sissel Tolaas. In *The Smell of Fear* (2006), Tolaas re-created, via chemical synthesis, the scent of men from different cultural backgrounds, all suffering from severe phobias. Tolaas then painted the colorless fluid onto a blank gallery wall, thus removing excess visual signification that might distract viewers. *Utter* differs from this groundbreaking work of Tolaas in its emphasis on live-ness, emergence, and performativity of microbes. It is central to this postanthropocentric approach that the scents produced in *Utter* (like *Labo*r) are *not* preprepared, but are produced on-site in a living, biological process. *Utter* and *Labor* projects seek to redefine the very verb "manufacture," as humans in this framework don't *facture*, but are *factured*.

The smells of human breath and body odors are generally abject, unwanted, and impolite. Much of Western culture has been designed and marketed to minimize such odors—deodorants, toothpastes, mouthwash, breath mints, laundry detergents, soaps, deodorizers, bathrooms, and the modern toilet. Historically, odor and myths of their effects have served racism, colonialism, imperialism, and classism well, reinforcing stereotypes, enacting boundaries, and dehumanizing perceived others. Janet Shaw, an eighteenth-century Scottish writer, wrote that in St Kitts slaves must serve water from a "cocoa nutshell ornamented with silver, at the end of a hickory handle" so that the breath of the slave "should not contaminate its purity" (Schaw 1922 in Tullett 2016: 315). In other Caribbean islands, such as Nevis, travel literature recounts that slaves were made to live in huts to the western side of the plantation so that slave owners could "breath the pure Eastern Air, without being offended with the least nauseous smell"(Sloane 1707 in Tullett: 2016: 315). In the nineteenth century, diseases such as cholera were surmised to arise from the odors of rot and decay in slums inhabited by impoverished and racialized others. It is this indeterminate, miasmic materiality of odor that has made it such an effective tool for racist mythologies.

Additionally, we are continually reminded of the odor of our own breath clinging to the pores of our masks, and others' oral odors have become a source of anxiety since they signal proximity and a possible viral transfer. The scents of others can arouse such fear as they transgress the very borders of self via a cloud of otherness, literally the material from another's body. However, it is hoped that audiences experiencing *Utter* find these artificially produced, vaguely human scents simultaneously comforting, reminding us of social, physical, and sensual human contact. In the profound Hawaiian concept of Aloha, *Alo* means to share while *ha* is breath, which is also life (Lee 2015). The exhaling of the syllable "ha" for the other to breathe in is to share one's life with another person. "Aloha" can mean a greeting, a welcome, or love, though English translations are inadequate as there is no comparable concept in the West (2015). Could it be possible to experience human scents as a welcoming transformation of the isolated self rather than as a hostile colonizing invasion?

By artificially recreating human oral scents, de-corporealized and in excess and offering them for aesthetic contemplation, I hope to befoul cultural prejudices through which both the faculty of smell and the quality of having a smell are denigrated. More importantly, I want to create a system for reimagining relations; a system that de-contextualizes the scent from its stigma; a network of apparatuses that de-index the sign from the body. The *Utter* project (its network of apparatuses) is not intended as an instantiation of an existing theory, but a relation within a phenomenon and hopefully an instigator of new theories.

Note

1 "Our" is intentionally ambiguous in this context: it could articulate ownership, but also collectivity.

References

Bake, B., Larsson, P., Ljungkvist, G., et al. (2019). "Exhaled Particles and Small Airways." *Respiratory Research* 20(8). https://doi.org/10.1186/s12931-019-0970-9

Bakhtin, M. (1981). *The Dialogic Imagination: Four Essays*. Ed. M. Holquist; trans. C. Emerson and M. Holquist. University of Texas Press.

Barad, K. (2003). "Posthumanist Performativity: Toward an Understanding of How Matter Comes to Matter." *Signs: Journal of Women in Culture and Society* 28(3): 801–31. https://doi.org/10.1086/345321

Braidotti, R. (2013). "Posthuman Humanities." *European Educational Research Journal* 12(1): 1–19.

Charles, M., Texier, F., Poinot, P., Arvisenet, G., Vigneau, E., Mehinagic, E., and Prost, C. (2013). "The 'Mouth to Nose Merging System': A Novel Approach to Study the Impact of Odour on Other sensory perceptions." *Food Quality and Preference* 28(1): 264–70. https://doi.org/10.1016/j.foodqual.2012.10.003

Dansinger, M. (2020). "What Is Acetone Breath, and Is It Tied to Diabetes?" *WebMD*. https://www.webmd.com/diabetes/acetone-breath-diabetes

Haraway, D. J. (2015). *The Companion Species Manifesto: Dogs, People, and Significant Otherness*. Prickly Paradigm Press.

Jay, M. (1993). *Downcast Eyes: The Denigration of Vision in Twentieth-Century French Thought*. University of California Press.

Jones, C. A., and Arning, B. (2006). *Sensorium: Embodied Experience, Technology, and Contemporary Art*. MIT Press.

Latour, B. (1999). *Pandora's Hope: Essays on the Reality of Science Studies*. Harvard University Press.

Lazzarato, M. (2009). "Mikhail Bakhtin's theory of the utterance." *On Bakhtin: video recording of a symposium*, Trondheim Academy of Fine Art. Trans. Arianna Bove. https://www.generation-online.org/p/fp_lazzarato6.htm

Lee, J. H., et al. (2015). "Asian American Religious Cultures." *ABC-CLIO*. ProQuest Ebook Central. http://ebookcentral.proquest.com/lib/buffalo/detail.action?docID=2145424.

Marchesi, J. (2014). *The Human Microbiota and Microbiome*. Cardiff University.
Murphy, H. (2020). "Surfaces? Sneezes? Sex? How the Coronavirus Can and Cannot Spread." *The New York Times*, May 26. https://www.nytimes.com/article/coronavirus-how-it-spreads.html
Rheinberger, H. J. (2019). "Epistemics and aesthetics of experimentation: Towards a hybrid heuristics?" In Sormani, P., Carbone, G., and Gisler, P. (eds.), *Practicing Art/Science: Experiments in an Emerging Field*, Routledge.
Robinson, A. (2011). "In Theory Bakhtin: Dialogism, Polyphony and Heteroglossia." *Ceasefire*. https://ceasefiremagazine.co.uk/in-theory-bakhtin-1/
Strecker, L., and Spiess, K. (2020). "Microbial Keywording: Towards material speech acts." *Performance Research* 25(3): 56–62. https://doi.org/10.1080/13528165.2020.1807755
Tang, G., Yip, H. K., Cutress, T. W., and Samaranayake, L. P. (2003). "Artificial mouth model systems and their contribution to caries research: a review." *J. Dent* 31(3): 161–71. doi:10.1016/s0300-5712(03)00009-5
Tullett, W. (2016). "Grease and Sweat: Race and Smell in Eighteenth-Century English Culture." *Cultural and Social History* 13(3): 307–22. doi:10.1080/14780038.2016.1202008
"Utter." OED|Oxford English Dictionary (2023). OED.com.
Vanouse, P. (2019). *Labor*. http://www.paulvanouse.com/labor.html
Whittle, C. L., Fakharzadeh, S., Eades, J., and Preti, G. (2007). "Human breath odors and their use in diagnosis." *Annals of the New York Academy of Sciences* 1098: 252–66. doi: 10.1196/annals.1384.011. PMID: 17435133
Xu, Z., et al. (2012). "Molecular and Microscopic Analysis of Bacteria and Viruses in Exhaled Breath Collected Using a Simple Impaction and Condensing Method." *PLoS One* 7(7): e41137. doi: 10.1371/journal.pone.0041137

18

Death Tool Kit

Practical and Spiritual Guidance from Artists Adriene Jenik, IONE, Marne Lucas, Linda Mary Montano, and Kira O'Reilly

Kathy High

Turn toward this thing called death. Even if you are afraid to turn toward it, turn toward it. It won't hurt you. And see what it has to teach you. It's a tremendous teacher, free of charge.

THE MIDNIGHT GOSPEL, WARD AND TRUSSELL, 2020

As the COVID-19 pandemic progressed with staggering numbers of deaths, in the United States we faced our deepest fears about mortality. People of color suffered the highest percentages of inpatient deaths (Vardar 2023). Harsh inequalities of race, gender, and economics were revealed, with tragic stories of people dying alone without enough care. In this essay, written during the lockdown of the pandemic, I consider the way that the arts and humanities can transform our denial of our own mortality. Practitioners around death care are essential for all populations, including Black, Brown, Indigenous, Queer, and Trans peoples. Through interviews with five artists working with death, I discuss ways to consider and honor death, to face our fears of dying, and to develop rituals for the dead and dying. This is the beginning of a larger project to create an artists' *Death Tool Kit*, providing strategies for being in conversation with death.

While clearing space for my home COVID-19 teaching office, I unearthed a small piece of paper with a list of instructions about how to prepare for death by Linda Mary Montano. Entitled "Death Notes," Montano wrote this list in 1988 on how to care for the dying (and oneself) as she had cared for her mother during the last five weeks of her life. It inspired me to write this essay, which, since the pandemic, holds new meaning for me.

During the pandemic, we have been exposed to over a million deaths in the United States and nearly 7 million worldwide at the time of writing,[1] mostly affecting people of color and those with little access to healthcare. In the United States, we are a "death-denying" nation—a culture that does not talk about death. However, the pandemic has forced our hand. As a result, I wanted to put together a *Death Tool Kit* that helps grapple with what might be called a "good death," crossing over, transforming to our next selves, struggling with grieving, and letting go.

To help me, I asked for input from five contemporary performance artists working with death. I interviewed them, incorporating their words along with my own thoughts as I mused upon what a *Death Tool Kit* might contain. What art and rituals do we need to help each other with our own deaths? Artists are particularly good at creatively reimagining the collective circumstances we find ourselves in, questioning and inventing new ways to consider and move through what we are given. I asked artists Adriene Jenik, IONE, Marne Lucas, Linda Mary Montano, and Kira O'Reilly to propose tools. Two are death doulas, and the other three are spiritual practitioners and counselors. All are teachers. I am grateful to them all for sharing their time, reflections, and wisdom with me.

Personal Account

In the television series, *Midnight Gospel*, real conversations are adapted from podcasts produced by comedian Duncan Trussell. Animated characters perform the texts with the help of trippy cartoons. The series's last episode, "Mouse of Silver" (2020), is a deeply personal conversation about dying between Trussell's comic character, Clancy, and Trussell's real mother, Deneen Fendig. In this episode, Deneen, taking on the character of Mother, speaks with Clancy (Trussell) after her real-life diagnosis of metastatic breast cancer. With soft fierceness, the characters both visibly age—Clancy from infant to adult, and Mother from young woman to elder. They nonchalantly continue their conversation. Mother says:

> People try to avoid the consideration that they are going to die and that people they love are going to die. It opens your heart. It breaks your heart open . . . If you inquire into the hurt, you know what you are experiencing

is love, the real deal . . . I am living and dying consciously. Simultaneously, I am holding both. ("Mouse of Silver" 2020)

As Mother moves closer toward death, Clancy asks, "What do you do about the heartache?" Mother responds, "You cry, you cry," and the characters hug and weep together.

I was introduced to death at the age of 25 when my mother died. Mom was diagnosed with end-stage colon-rectal cancer, and my family agreed to support her wish to die at home. My father and I cared for her during her last seven months of life, along with periodic visits from my two brothers. We were all present as my mother passed. I held her hand and watched her die, as my older brother held my other hand. It was shocking and hard to witness. One could say my mother had a "good death." Although even now, forty years later, it is hard for me not to also remember how painful it was to watch her die. I wished Mom had more pain management. But "good" in the sense that we and her primary doctor respected her wishes to die at home, and "good" in the sense that my mother waited for our entire family before she passed, holding her hand and witnessing her passage. My siblings and I were also with my father when he passed away in the hospital fifteen years later. This shared presence with both my parents as they died made my grieving easier. I would do things differently now: I engaged in "positive" visualizations with my mother imagining her healthy cells overcoming her cancer cells. After months of this exercise, my mother burst into tears crying that she had failed me for not being able to defeat her cancer. Suddenly, I realized that my prompts to her were to live, and I was not allowing her to process her impending death. I wished I had visited my father more the last month of his life when, in his demented state, he only recognized me as my mother. After their deaths, my brothers and I assured each other we had done our best. My parents' deaths broke my heart open.

The Midnight Gospel episode moved me to tears as I witnessed Clancy and Mother being so honest with each other. I have watched it repeatedly, hoping it will teach me to do likewise. This didn't happen in our family. We rarely spoke about emotions—particularly negative ones. The speed of the pandemic stole time from families dealing with the emotional toll of death. We need more tools to imagine the end of life and to accept our mortality.

Death Tool Kit is partly conceptual but also practical. I have thought of it much like *Boîte-en-valise* (1935–41), the suitcase project that Marcel Duchamp developed as a kind of retrospective of his own career, containing miniatures of his artworks. *Death Tool Kit* could be a collection of works and radical strategies by and for those working with death and impermanence. Conceiving it has helped me to understand death as a process, where the body can decompose and transform to another beautiful material. Not simply an ending as medical science and statistics would suggest, but a new way to be in the world.

Practicing Death

Since my mother's death, I have been curious about decay, decomposition, and the processes of dying and transformation. Shortly after my mother passed in the 1980s, I moved to New York City and was part of the AIDS crisis there. I participated in ACT UP, DIVA TV, and other tangential activist media groups fighting for AIDS recognition and treatment. Many dear friends passed, and I tried to be there for them to let them speak about their illnesses and pending death.

As an artist, imagining my own death or the death of others as ritualized performance has allowed me to experience death bit by bit, to become familiar with the idea of losing my own precious life—"practicing death" as a kind of self-care, rather than negating it. For example, I had a superstition that I was going to die in my forty-fifth year, the year 2000. I created rituals around this fear to help me greet my own mortality in *Everyday Problems of the Living* (Figure 18.1). Each month, I performed and filmed my own

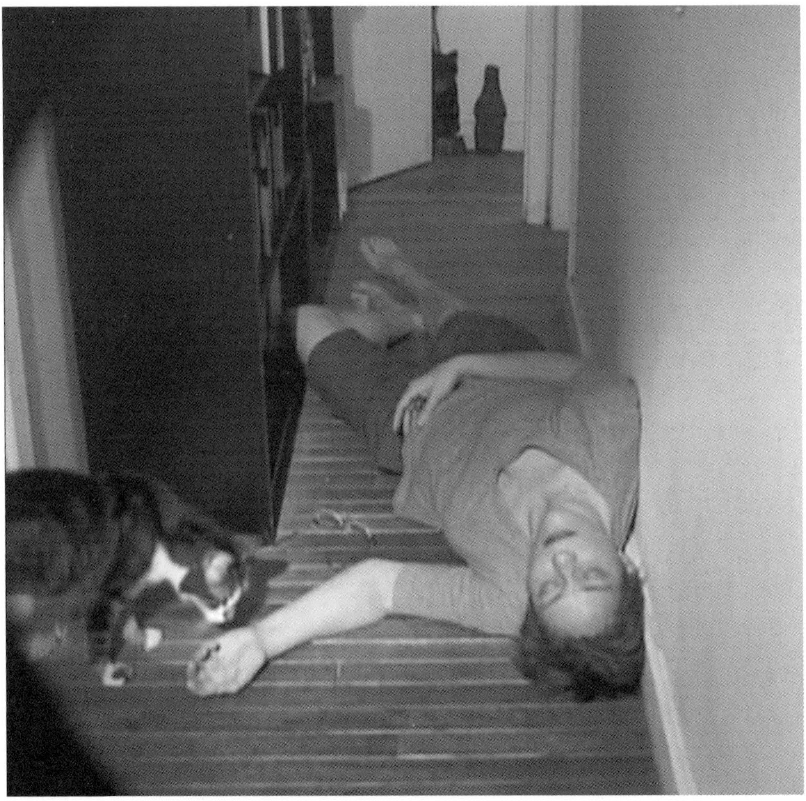

FIGURE 18.1 *Kathy High,* Everyday Problems of the Living: Death Poses, *2000. Courtesy of the artist.*

death, or a ritual around my death. Sometimes, I literally pictured ways I could die: falling down a stairwell while my dog barked; expiring in my kitchen holding a spatula; perishing inside my car going through a carwash. Often my pets would comically interfere, interrupting my performances. I found the productions to be cathartic and allowed self-mourning to happen slowly over the year. And, as you might surmise, I did not die that year—my superstition was incorrect.

I have continued to "practice death." For example, I purchased a green burial spot for myself in a natural cemetery. I have been designing a garden to be seeded on my grave. During the pandemic, this kind of burial became a luxury, one not afforded to the rising numbers of the dead, unable to find plots to accommodate the bodies. Perhaps moving to human composting and establishing facilities to decompose bodies makes more sense as we look to future pandemics.

Loving One's Fear of Dying

Each of the women interviewed here is an artist and practitioner of rituals around death. Since the 1970s, Linda Mary Montano practices what she calls a living art. In our personal interview on January 21, 2021, she stated: "I use art as medicine, I use art as therapy, I use art as spiritual practice. And art now uses me. I call it Lifeist . . .which is an artist who makes art of life." Montano spoke with me about her performance of death throughout her career. This obsession with death began as a result of "repressed internalized trauma" from when she almost died very young. One of the first artworks of Montano's I encountered was her 1977 video *Mitchell's Death* where she mourns her ex-husband's death. "Every detail of her story, from the telephone call announcing the tragedy, to visiting the body, is chanted by Montano as her face, pierced by acupuncture needles, slowly comes into focus then goes out again" (Montano *Video Data Bank* 1977). Montano learns lessons of forgiveness and "loving one's fear of dying" along the way.

Montano's *Death Notes* is a poignant and frank list of advice written while sitting with her dying mother. What are the small things we learn from witnessing a loved one passing? What pragmatic lessons do we want to pass on to others? Some of Linda's notes include: "Help the way they need to be helped, not the way you need" and "See if they have unfinished business they want to work with." "Whisper messages near her ear." "Listen, listen, listen." Her beautiful advice about her mother's death and being *with* her mother at the end is a touching summary, ending with "Prepare for your own death in your own way" (Montano 1988).

Linda M. Montano

DEATH NOTES

In 1988 my mother died. It was a five week hospital death. Colon cancer. I was with her much of that time. These are some of the things that I learned.

1. Ask the dying person questions and wait for verbal/ nonverbal responses i.e. "Should we stop this medicine?" "What do you need?": (I didn't always ask my mother's opinion and now wish I had).
2. Listen, listen, listen.
3. See if they have unfinished business they want to work with, i.e. unexpressed thoughts, feelings, wishes, etc. (My mother said, "I always wished I had written a book," knowing that I would hear her and do it for her. I have.)
4. If they prefer, let them leave consciously, alertly, with awareness. (Because one medicine was stopped, she became comatose. I didn't know that would happen and I missed saying goodbye because of that.)
5. If they agree, play nature tapes in the hospital room. It creates a no panic atmosphere. (I played one and nurses used to come in the room because it felt "good in here").
6. Help the way they need to be helped, not the way you need. (Once I was chanting, praying, laying my hands on her abdomen, teaching her to breathe and my mother, always the comedienne, opened her dying eyes, looked at me with an Imogene Coco-look and said, "Linda, please!")
7. Confess by their bedside. (Clear your heart but do it telepathically. Tell everything you need to tell. Do unfinished business without their hearing. Sotto voce.)
8. If the person is on heavy painkillers, they might change their behavior toward you, positively or negatively. Be ready! (My mother became a hippie on morphine. She pulled me close to her, tried to show me my aura, touched the peach fuzz on my cheeks. We re-bonded.)
9. Get counseling yourself from either friends, 24-hour telephone hotlines, Hospice volunteers, etc. I talked with a Hospice volunteer for 4 years on the phone while my mother was sick and dying. (Hurrah for Hospice!!)
10. Know what "patient's rights" means and use the information appropriately. (I asked one nurse, "If this were your mother would you allow a nurse to take blood and do "vitals" while your mother was comatose and had only a little time left?" The nurse said, "No." I said, "Then please do not take blood and stop all orders at the desk.")
11. Be prepared for each family member and close friend of the dying person to have a completely different response to everything. Emotions are close to the surface; everyone's death anxiety button is being pushed.
12. Use TV, VCR's as teaching tools. (When my mother wanted TV on I turned to cartoons. The flying horse image from My Little Pony did wonders for her attitude and actually kept her pain-free.)
13. Inconspicuously breathe together (match your breathing pattern to theirs), and gently sound the exhalation. (Steven Levine teaches this technique. It's Tibetan and comforting for both patient and caretaker.)
14. Whisper messages near an ear. Keep messages positive and in words they need to hear. (When I said, "Relax, Ma, Mitchell and Karl will take very good care of you," she responded positively. These were friends who were dead and whose company she loved.)
15. Go as far as you can go with the process at the end. (Those last 15 hours when she was comatose, I felt a need to meditate, give her space, not touch her as much. Check out your own situation, it will be different).
16. After the death participate in the process as much as you can given place,circumstances, etc. Help wash the body, close the eyes, etc. (I didn't see the spirit or soul leave. Some people do.)
17. Never, ever judge how you or anyone else mourns or deals with a death. (I demolished and rebuilt an abandoned building for two years after she died. I rarely cried, but my body sweated.)
18. Be watchful for messages, dreams, symbolic visits. (She comes as a butterfly or sometimes a wave of feeling. I continue a dialogue with her.)
19. Daily: Prepare for your death in your own way.
20. Your comments:

Typesetting: Janet Bizzell

FIGURE 18.2 *Linda Mary Montano,* Death Notes, *1988. Image courtesy of the artist.*

Assisted Dying

Artists Marne Lucas and Adriene Jenik are certified end-of-life doulas. They assist the dying and their families through the death process, explaining what to expect physically when someone dies, negotiating wills, arranging doctor's visits, and helping create rituals for passing. Both spoke about the need to be adaptive to the dynamics of families they serve, respectful of their needs and traditions. Lucas and Jenik act as mediators who honor the wishes of the dying. Deep listening is key—observing everything from spoken communication to body language and subtle spiritual signals. In our interview, Lucas outlined her work:

> [W]e're there to spiritually assist the patient and their family and loved ones to a place of acceptance and ego integrity. But part of what a doula does is convey the wishes of the dying. The dying don't get a lot of sense of personal agency. So, I think a lot of dying people end up feeling they're left out of the conversation about their own death. My role is to have those uncomfortable conversations with that dying person: What do you think your death is going to be like? What do you want to happen aesthetically, spiritually, emotionally, physically? How do you want to be honored posthumously?

They spoke of the hardships of working with people during COVID-19 and how people died isolated and alone. Lucas reflected on our collective protracted grieving from a lack of physical presence, noting how strange it was to work on something as intimate as death without being able to touch each other. For Jenik, witnessing the influence of the AIDS crisis on her queer community in the 1980s shaped her, teaching her to be present with death and the dying. From that experience she knew she could be a doula. Queer communities, like others, need protected space to feel safe when honoring the dead and dying. Jenik likened the COVID-19 relationship with collective and digital activism, mutual aid, and Black Lives Matter to the organizing around AIDS by ACT UP where "some people are transforming their anger around all this needless death into activism" (Jenik 2021).

Both have worked on projects around dying that are life-affirming and useful. Lucas created the *BARDO ∞ PROJECT*, a project for artists who are dying or have a terminal diagnosis to prepare artwork for their legacy. Lucas's first project was with her best friend, photographer Chris Brunkhart (1967–2016), who had metastatic colon cancer. She helped Brunkhart through his process of dying, supporting the mourning of partner/husband, Zeke. Lucas helped produce Brunkhart's last body of photographs, "Borderless Space" (2015), documenting "Chris's experiences while living in New York City" (Lucas 2021). Brunkhart died ten days after his exhibition opening and book signing. Lucas continues the *BARDO ∞ PROJECT*, working closely with dying artists and their families to identify and bring to fruition their final works. "It's not about the kind of darkness of death and dying. It's really to celebrate that person's life. Ultimately it is an optimistic project" (Lucas 2021).

Jenik created a deck of *Eco Tarot* cards that predict our environmental futures and consider mass extinction and climate crisis. From this project she oversaw the creation of *The Artist Grief Deck* during COVID-19, a deck of cards with artwork by artists paired with different statements around grieving. Because of her death doula work, Jenik participated in the artists working group on collective grieving. She worked with NYC VOAD (New York City Voluntary Organizations Active in Disasters):

> [T]hey identified this unmet need for assistance with collective mourning, or individual mourning, because people weren't able to access their faith community, or access each other. People were stranded, and there was this very challenging situation, and so they thought that maybe an artist working group could think of unconventional ideas. (Jenik 2021)

Jenik proposed the idea for *The Artist Grief Deck*, which was realized with multiple artists' input. These art projects aid in the ability to envision one's wishes for the dying, and to help those of us left behind to process our grief.

The Diversity of Grief

Issues of inequity around race and class were also key to Lucas's and Jenik's observations around the pandemic. Both recognized the severe inequities that were revealed. The activism and civil unrest around Black Lives Matter also spoke to what Lucas described as a social "Black grief" stemming from long-term oppression and systemic racism. Lucas said she would like to see "people be focused on being in the service of others, as opposed to just devout capitalism" (2021). Lucas started to host end-of-life doula meetups prior to the pandemic's lockdown. She noted that participation in these meetups was primarily white and she expressed a deep commitment to diversify these meetings. Lucas, a white woman, has been in touch with many end-of-life doulas of color and felt it imperative to integrate and learn from them. No one doula can serve all communities, and so Black, Brown, Indigenous, Queer, and Trans doulas are essential. Jenik applauded the promising growth of the end-of-life doula profession, but pointed out that members of the Black community or others of color may have already been practicing end-of-life rituals. She hoped that professionalization would not cut them out or negate their traditional knowledge—just as midwives were sidelined by the male-dominated American Medical Association in the nineteenth century, limiting forms of healing knowledge (High 1989).

The women in this essay challenge traditions, upturn power dynamics, and offer rituals and emotional support to those they comfort and assist. They show us how to be present, respect the process, and listen deeply. For example, Jenik has observed the Buddhist practice of Tonglen with people who are dying, a meditation practice where one breathes in suffering and breathes out peace, love, and comfort, to aid those who are passing. Jenik said, "It is a beautiful practice that brought a community together at the end of my friend Lesley's life (even virtually). She passed from cancer in January but was almost completely pain-free (a complete miracle from my perspective working with other cancer deaths) and we all liked to ascribe that to our dedicated Tonglen practice" (2021).

Presence Without the Body

IONE was the long-time partner and collaborator of the late electronic music composer Pauline Oliveros. Together they worked tirelessly with the Center for Deep Listening, teaching the practice of Deep Listening pioneered by Oliveros in the early 1970s (Oliveros 2005). In the 1980s, IONE founded the Ministry of Maåt, Inc., to promote global harmony and women's wellbeing. IONE is a playwright and experimental theater director, most recently *The Nubian Word for Flowers: A Phantom Opera*, cocreated with Oliveros (2017).

She authored *Pride of Family; Four Generations of American Women of Color* (IONE 2007), tracing her African American matriarchal lineage from the 1860s onward. She is a counselor, teacher, and support coach focusing on ancestral connections, dream work, and nurturing our own creative spirit.

IONE understands these deep histories and legacies from our past. When we spoke about the pandemic, IONE described our collective grieving as not unique to COVID-19, but rather ongoing. Because of our culture's engagement in constant wars, and even violence promoted by religion, we are in a constant state of mourning. IONE said, "I think that it was the Dalai Lama who said . . . and this was during one of our wartime situations . . . that we should make war illegal. *Make war illegal.* And I thought, yes! Yeah, that's it! . . . So, I would suggest that we make war, and even war-like gods illegal, or do away with them, one way or another in some harmonious way. (*Laughter*)." This war-state perpetuates a constant cultural grief.

In terms of the ever-present threat of death during the COVID-19 pandemic, IONE commented:

> It's both confrontational, and also has a healing element to it, perhaps, that was quite profound. People are recognizing: "oh okay, this is it!" And being unable, really, to be as unconscious about life itself as they

usually are. So, this is a kind of enforced study of one's own existence, which is very disturbing, because you can't get away from it in the same way that you could before, without being brought back to: "oh there's a pandemic, and I might get it and die" . . . I noticed that with so many people . . . everything's up and you have to pay attention to what is, shall we say, which we spend our lives trying to get away from. (*Laughter*). (IONE 2021)

The global impact of COVID-19 could not be avoided. Because of it, we have recognized the thing that some call the "now," being present, or as IONE said, what *is*, including what we "not-do" rather than "impose on others, that allows the process to unfold into a person, to feel a support" (IONE 2021).

COVID-19 has been a teacher of death. Families have been denied physical access to each other as loved ones were sick and dying: the sick isolated in ICUs or hospital wards often saying final goodbyes to family over video calls. Montano referred to COVID-19 as "leveling the playing field," where she felt an encompassing empathy: "I am one with everyone." While it is important to acknowledge that grief is personal, Montano suggests many were "spanked by the pandemic into, 'Let go, it's all dying, let go and just die. And be invisible.'" While people were dying alone in hospitals without family to hold them, Montano felt one lesson was surfacing from being present: "[B]ecause we can't attend the bodies of those close to us, we are being forced into a very, very high spiritual position of learning how to be present without the body. And learning how to work on telepathic channels, on angelic levels, on imaginative levels, on electromagnetic levels," adding "we're going to understand the love of the fear of dying."

From Wakefulness into Dissolution: Trusting the Bardo

Artist Kira O'Reilly works to create a body consciousness through her practice. She is a visual artist who works with her body in performances to think through questions of time, life, death, and embodiment. Kira works on her own physicality to "really allow for working with having a body: what is it to be embodied? what is it to be made of materiality?" In her endurance performance *Stair Falling* (O'Reilly 2010), O'Reilly slowly "falls," descending an elegant staircase, naked, doing a kind of slow-motion downward backwards somersault. This work conjures the slow trauma we experience through death and life, a visceral reference to the miracle of our most intimate moments and endurance. Recently O'Reilly developed artwork and performance around the death of her close friend, Darren George. She stated she has been "working with material from Darren's death, like the autopsy, finding ways to work with it that aren't sensational, and aren't explicit." O'Reilly continued that she has been considering their death through written texts and her own breath: "I have to do [the performance] with my breath, and my body, the autopsy is read on an inbreath and amplified—after the inhalation the reading continues so the sounds are only of the interior of my mouth, empty and resonant somehow. It's just really using the kind of technology of the mouth as something that amplifies. The body being referred to exceeds the autopsy just as the autopsy evacuates the body" (O'Reilly 2021).

O'Reilly is ordained in a Tantric Yogic tradition of Vajrayana Buddhism, practicing in the Nyingma Aro gTér Lineage, which includes practices in relation to death. During the pandemic, O'Reilly spoke about holding others, but at a distance, performing a spiritual holding of others as they pass from this reality. As a Buddhist, she practices with people both as they approach death and afterwards. She also compared the trauma we culturally experienced from COVID-19 to that of the AIDS crisis, a collective grief. Through this spiritual embrace and her practice for people as they die, she created the feeling that "someone's got my back," when we could not be physically present for each other, supporting transition through the Bardo. She stated:

> . . . to feel that someone's got my back, not only as I'm feeling like shit from the side effects of illness, but as I actually dissolve in the death process itself and enter into the Bardos. Bardo is a word that is used in Tibetan Buddhism for this period after death. Bardos aren't just about the death process. One of the experiences we have every day that is almost the same as death is falling asleep, the process of dissolving, of going from wakefulness into dissolution. (O'Reilly 2021)

IONE also spoke of the Bardo, "Teachings of the Bardo are about us coming to terms with that fact that it's not only after death, it is not only about learning how to move through after death, it's now, it's about now" (IONE 2021). Belief in an afterlife and possible rebirth has helped me understand the importance of not leaving things in my life "undone," but to find resolution even to the most painful relationships and events.

IONE and O'Reilly pointed to our periods of sleep and dreams as active ways to prepare us for death. IONE called dreaming the "sneaky revolution" where we see our lives unfiltered with full awareness, an awakening to the real. "Because what happens when people begin to pay attention—or have more awareness—in their dreams are getting closer to those states" (IONE 2021) of enlightenment. Each artist addressed this reawakening and presence. This process was referred to as a *transformation*, a *regeneration*, a *vibrational frequency*, a *dissolution*, *rebirth*, and *unfolding*, a *surrendering* and *letting go*. Energetically, we circle back to love, unconditional love, that we all need to thrive in life and death. As Montano said, "there is no correct way to do it." It is "beyond good and bad, beyond ethics" (Montano 2021).

Conclusion and Dissolution

I have dealt with my own chronic illness for over four decades, so notions of mortality, fragility, and disability have been familiar territory for me. I have worked outside our standard medical system to seek guidance and health support. I have a deep belief in the powers of telepathy, psychic abilities, and mediumship. I have consciously been working with and learning from living systems, ranging from transgenic rats to other animal companions, immune cells, and gut fecal microbiota. I have stumbled through these biological materialities, tried to communicate as well as I can, and been struck by my own fears over again, while experiencing great humility and laughter. This is why I am looking at death and gathering tools for the journey. Now more than ever, I want to hone my *Death Tool Kit* so I can help myself and my queer, feminist, spiritual, artist communities, and chosen family.

I looked to the artists in this essay for guidance because each is a friend, a mentor, and part of my extended family. Their courage to face death head-on is what I want to glean and pass on. Artists seem well-suited for doing "death work." Jenik notes that we artists are "resourceful and creative," are "often coming up with rituals, and also we're improvisational." She mentions that there is "some kind of connection to spirit where we hone-in [while] downloading ideas or connecting" (Jenik 2021). This open practice allows for acknowledgment of our feelings, being present, and observant. O'Reilly asks us to consider that we experience dying in small ways all the time: "Death and life and rebirth are experiences that flicker through every minuscule moment of one's life and one can practice that experience." It can be a collective, communal ritual, dancing, and practicing life and death.

My neighbor and friend Paul, aged 88, died summer, 2021. He was the rock of my small community, generous with advice, wry humor, and nature stories. His loss made me consider his gifts. Paul taught me how to keep honeybees, keep *with* the honeybees as best we could. When I first worked with the bees, I was terrified by their sheer numbers. I learned to talk with them, and distinguish the pitch of their buzzing versus the yellow jackets who waited in the wings when we opened the hives. After Paul's funeral, his grandson asked if we could collectively keep some hives. This shared ritual of tending the bees is now the reanimation of Paul. This is part of my *Death Tool Kit*.

In *Midnight Gospel*'s last episode, the character of Clancy's Mother says: "[T]he amazing thing about letting go into it is that you find that this thing called love is supportive, it holds us" ("Mouse of Silver" 2020). During the pandemic, we witnessed worldwide catastrophic loss of life. But this project gave me the will to reanimate those we have loved and lost. The *Death Tool Kit* remains open-ended and needs more collective creative input. It offers amazing tools from the wise artists I interviewed: be conscious, adaptive, attentive, supportive, loving, creative, forthcoming, vulnerable, and unrestrained. I am designing the *Death Tool Kit* website, deathtoolkit.net, to house projects and ideas by this diverse community which is growing and will seed the project. This will be a repository for artists' mourning rituals around extinction, tools of love around end-of-life, and small ways to start letting go. Each of us must become death doulas for ourselves and develop this tool kit for our families, the planet, and its creatures. Death is our great teacher.

Note

1 There have been at least 1,129,838 Covid-19 deaths in United States (CDC 2023) and 6,935,889 deaths worldwide (WHO 2023) as of May 23, 2023.

References

Aro Encyclopaedia (1997). *The Nine Bardos of the Aro gTér*. https://www.aroencyclopaedia.org/shared/text/n/nine_bardos_ar_eng.php

Centers for Disease Control and Prevention (n.d.). https://www.cdc.gov/coronavirus/2019-ncov/covid-data/covidview/index.html

High, K. (1989). *I Need Your Full Cooperation*. Video. https://www.kathyhigh.com/videos/i-need-your-full-cooperation/

High, K. (2005). *Everyday Problems of the Living*. Video. https://www.kathyhigh.com/videos/everyday-problems-of-the-living/

High, K. (2015). *Burial Garden: Becoming a Plant: Resurrection Garden*. Video. https://www.kathyhigh.com/projects/burial-garden/

High, K. (2021). Interview with IONE, conducted January 30, 2021.

High, K. (2021). Interview with Adriene Jenik, conducted February 13, 2021.

High, K. (2021). Interview with Marne Lucas, conducted January 20, 2021.

High, K. (2021). Interview with Linda Mary Montano, conducted March 31, 2021.

High, K. (2021). Interview with Kira O'Reilly, conducted January 26, 2021.

IONE, C. (1991). *Pride of Family, Four Generations of American Women of Color*. Broadway Books.

IONE, C. (1997). *The Ministry of Maåt*. https://www.ministryofmaat.org

IONE, C., and Oliveros, P. (2017). *The Nubian Word for Flowers*. Opera. https://experimentsinopera.com/portfolio-item/nubian-word-for-flowers/

Jenik, A. (2017). *Eco Tarot*. Tarot Deck. https://theconnectcenter.asu.edu/ecotarot-deck

Jenik, A., and The Artists' Literacies Institute (2020). *The Artist Grief Deck*. https://griefdeck.com/

Lucas, M. (2021). *BARDO ∞ PROJECT*. Various media. https://www.marnelucas.com/bardo-project-home-page

Montano, L. M. (1977). *Mitchells' Death*. Film. United States: Video Data Bank. https://www.vdb.org/titles/mitchells-death

Montano, L. M. (1988). *Death Notes*. Handwritten Instructions.

Montano, L. M. Video Data Bank. "Linda Mary Montano Mitchell's Death." https://www.vdb.org/titles/mitchells-death

Oliveros, P. (2005). *Deep Listening: A Composer's Sound Practice*. iUniverse.
O'Reilly, K. (2010). *Stair Falling*. Performance. http://www.kiraoreilly.com/stair-falling
The Center for Deep Listening (2014). https://www.deeplistening.rpi.edu/
The Midnight Gospel (2020). S1: E8 *Mouse of Silver*. TV Program. Netflix series.
Vardar, U., Ilelaboye, A., Murthi, M., Atluri, R., Yong Park, D., Khamooshi, P., Ojemolon, P. E., and Shaka, H. (2023). "Racial Disparities in Patients With COVID-19 Infection: A National Inpatient Sample Analysis." *Cureus* 15(2): e35039. https://doi.org/10.7759/cureus.35039
World Health Organization Coronavirus Dashboard (n.d.). https://covid19.who.int

19

Creating and Exhibiting Artworks Embedded with SARS-CoV-2 Genetic Material during the COVID-19 Pandemic

Anna Dumitriu

Introduction

As I write this in June 2021, I am able to reflect on the past year and a half and begin to describe my experience of the COVID-19 pandemic so far. As an artist whose work has focused on microbiology and the impact of infectious disease for more than twenty years, I found myself watching a phenomenon that I had studied for much of my life—the world-changing impact of infectious disease—play out before my eyes.

During the pandemic I created a series of works which incorporate a safe, noninfectious version of the SARS-CoV-2 RNA (coronavirus), supplied by collaborators Ines Moura and Jane Freeman at the University of Leeds, UK.[1] These bioart works allow us to engage with the virus, which killed over 20 million people [estimated global excess deaths], without risking contagion. They express ethical, aesthetic, and safety concerns around the virus, interrogate how the virus impacted humanity, and explore health and social inequalities.

My sculptural and installation-based works often draw threads across time from the history of science and medicine to cutting-edge research in synthetic biology and genomics. In this text, I describe two recent artworks, *Shielding* (Dumitriu 2020) and *Lockdown Votives* (Dumitriu 2020), that explore the history of infectious disease and antibiotics, and the collateral effects of the COVID-19 pandemic and conclude by reflecting on the laboratory relic as sublime.

There is a Sense that the World is Heading Toward a New Pandemic . . .

When asked that classic dinner party question, "If you could go back in time, when and where would you go?" I would always say, "the Great Plague of London, but with a vaccine of course." Having worked extensively with *Yersinia pestis* (the bacterium that causes the plague) in numerous projects, I was fascinated in its impact on human behavior and how even now we are still finding out new information on how it spread (predominantly through human body-lice rather than fleas).

In 2012, I wrote:

> There is a sense that the world is heading toward a new pandemic, that an unknown disease will emerge or that an existing pathogen will evolve strategies to resist our limited antibiotic cures and strike us down. At the heart of this a sense of the sublime, combining terror and awe as we reflect on the impact of these minute "life forms," whose complex behaviors we are only now beginning to understand. New developments in whole genome sequencing of microbes now offer us hope, potentially enabling doctors to diagnose and precisely treat diseases in a matter of hours. But the implementation of whole genome sequencing (WGS) technology in microbiology raises a number of complex technical and ethical issues from processing and understanding vast amounts of data to potential privacy issues revealed by the specific organisms an individual may be carrying and passing on, and even in the way genomic data is pieced together jigsaw puzzle-like. The cost of WGS technology is continually being reduced while processing speed is increasing exponentially at an unexpected rate, and it will begin to be implemented across healthcare providers worldwide within the next five to ten years. We are in the midst of a quiet revolution that may have as big an impact on our lives as the industrial revolution had on our ancestors' lives. (Dumitriu 2012)

The Plague Year(s)

On February 6, 2020, in my hometown of Brighton (UK), I found myself in the unenviable position of living in the city where the first British case of COVID-19 emerged. By March 23, 2020, I found myself locked down in the UK, living through this modern-day "plague" and trying to comprehend the striking similarities between my everyday life and the experiences of great diarist Samuel Pepys ([1633–1703] 1932), or written about by Daniel Defoe in *A Journal of the Plague Year* ([1661?–1731] 2003), texts which had fascinated me since childhood.

Spring 2020 was a period of upheaval as my plans were thrown into chaos but slowly the first shoots of the "new normal" began to emerge, and with that, I developed new methodologies and strategies referencing the pandemic, in order to reveal hidden stories.

Shielding

In late 2019, I was commissioned to cocreate an artwork with staff from a charity working with survivors of domestic violence. Just before beginning a series of workshops, which I had planned to develop the work, everything had to be canceled. I had to find a new way to produce work on the theme, alone. It was not long before news stories from around the world started to highlight the significant upsurge in domestic violence. There were calls for increased support for victims of abuse from both governments and the charity sector, at a time when support was much harder to access because of infection control measures and reduced capacity, especially in the early stages of the pandemic.

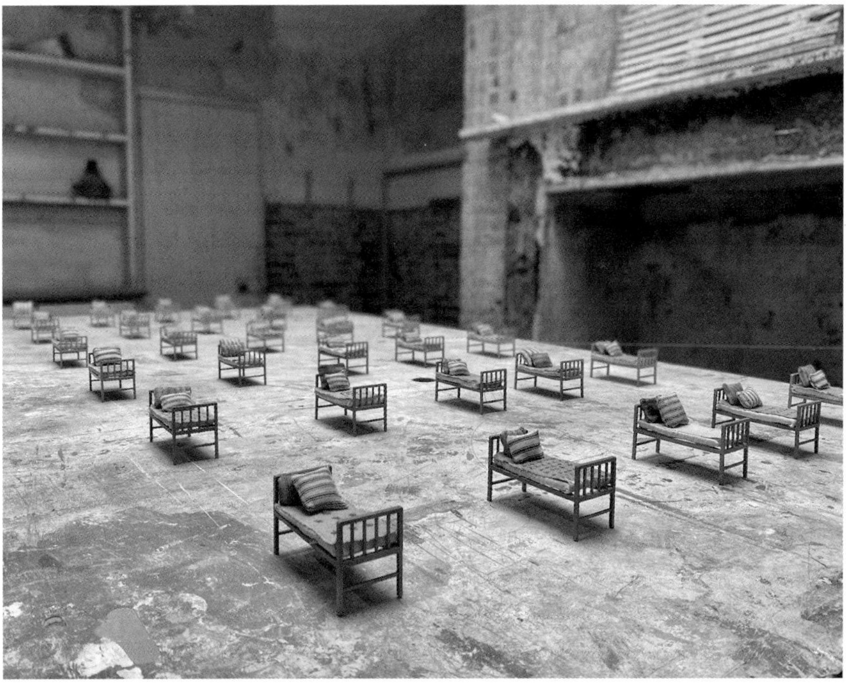

FIGURE 19.1 *Anna Dumitriu,* Shielding, *2020, installation view at the Regency Town House, Brighton, UK. Commissioned by ART/DATA/HEALTH. Photograph by Anna Dumitriu. Image courtesy of the artist.*

I created *Shielding*[2] in response to the data I collected from international press and TV news media. *Shielding* explores the impact of the COVID-19 quarantine on women facing domestic abuse and the paradoxical meaning of home as shelter. Ironically, the title comes from the term the UK Government used to describe the process of protecting those who are clinically vulnerable to COVID-19 who are asked to stay home (known as "sheltering in place" in the USA), but the reality is for those at risk of domestic abuse the home is not a place of safety and care, and helplines globally are facing an increase of up to one third (Oppenheim 2020).

The artwork is inspired by Virginia Woolf's feminist essay "A Room of One's Own" (1929), which states that women need their own safe space in order to flourish and be creative. This notion of a safe space is set against the stark image of the hastily constructed temporary hospital ward that has become such a familiar image in the news stories of 2020. Those locked down with abusive partners have no safe space in which to escape. Movement restrictions aimed to stop the spread of COVID-19 are making violence in homes more frequent, more severe, and more dangerous.

The installation combines traditional feminine crafts such as sewing, embroidery, and natural dyeing using healing plants (such as turmeric), with 3D printing based on digital reconstructions of hospital beds from the first temporary hospitals in Wuhan with their rows of identical beds, often with a bizarre jumble of colorful makeshift bedding sourced at speed from nearby factories. The doll-sized beds may also remind us of Henrik Ibsen's play *A Doll's House* ([1828–1906] 1972), which deals with the fate of a married woman in a male-dominated world.

Incorporating SARS-CoV-2 RNA from a Plasmid Construct

It is a frequent theme in my work is to incorporate actual genetic material or other materials from the laboratory process, such as in my *Plague Dress* (Dumitriu 2018), so I had the idea to try to safely incorporate biomaterial related to COVID-19 into *Shielding*. Even though my laboratory work and scientific collaborations were heavily disrupted by the pandemic "lockdowns," I thought it might be possible to take SARS-CoV-2 RNA and use it in the work, after all the RNA itself is not infectious. In fact, I had already discussed my idea of using it in an artwork with another team of collaborators in early March 2020. Subsequently, I approached my collaborator Jane Freeman who had started to work on COVID-19 research. Her colleague Ines Moura was able to send me SARS-CoV-2 genetic material. For additional safety, this RNA did not come from the real virus, but from a plasmid obtained from the National Institute for Biological Standards and Control. The plasmid is a piece of existing RNA where its genes are replaced by those of SARS-CoV-2 so that research can be done without any risk of causing disease. An important part of my work is to always check the health and safety of such materials and to collate the necessary documentation, including letters from scientists and institutions on headed paper, which allows the work to be exhibited, and this is all in place for *Shielding*.

I impregnated the embroidered beds with the SARS-CoV-2 RNA (coronavirus) given to me by Moura and Freeman. The presence of this biomaterial, even in a safe noninfectious form, allows us to confront some of our feelings about the pandemic, while the artwork itself becomes an authentic secular relic of the sublime laboratory and the research processes that we are still pinning our hopes on.

Since the creative workshops we had planned were canceled due to the pandemic, I prepared an online art kit for participants (staff working with domestic abuse survivors) instead. The creative activity prompted them to imagine an ideal, a room of their own or a safe and cozy space for a service user, reflecting on Woolf's writings. I hope that in due course I can run a physical workshop with participants such as staff or domestic violence survivors to enrich the work by creating miniature bedcovers with embroideries that bring their own meaning to the piece.

The exhibition was installed in the old kitchen space at the Regency Town House in Brighton in October 2020 and fully documented, but due to health and safety restrictions placed by the funders in terms of running public events during the pandemic period, it could not be physically experienced by audiences. It was finally launched via an online conversation event as part of the Brighton Fringe Festival on October 9, 2020. Since then, I have shared the work through multiple online talks and I look forward to physically showing the work in due course.

Lockdown Votives

I started work on *Ex Voto* in 2016. It is an ongoing participatory artwork and installation that explores the impact of infectious diseases and antibiotics on our lives, created through the making of "votive offerings" during drop-in story-sharing workshop sessions. These secular votive offerings reference those found in religious settings symbolizing a wish or giving thanks for its fulfilment. The votives are hung on ribbons, stained or dyed with bacteria, including various species of gut microbiota, the *Staphylococcus aureus* from my own body, and modified antibiotic-producing Streptomyces (all sterilized), as well as natural antimicrobial substances such as madder root, and nonhazardous chemical dyes used in the lab.

I extended this project during the COVID-19 pandemic, and in particular the first UK lockdown period in 2020. I documented my experiences and combined them with reflections and memories of the period that I heard about through friends, colleagues, social media, and the news. The *Lockdown*

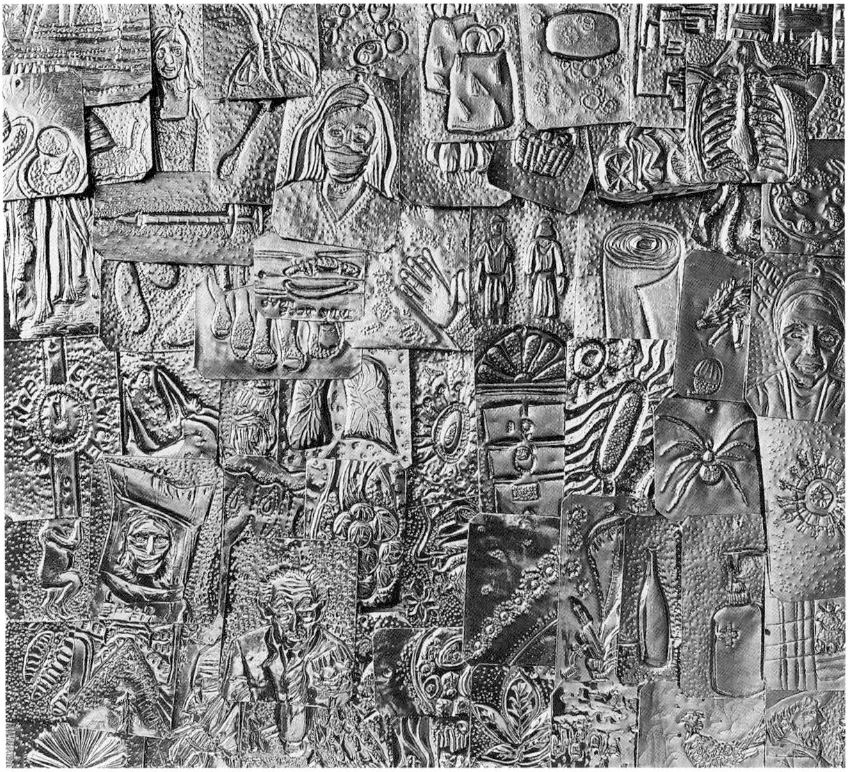

FIGURE 19.2 *Anna Dumitriu,* Lockdown Votives, *2020. Photograph by Anna Dumitriu. Image courtesy of the artist.*

Votives series explores the changing face of treatment of the disease, recovery stories, and other remembrances of the period. These votives are hung on turmeric-dyed ribbon. Turmeric has been a popular remedy during the pandemic due to potential anti-inflammatory elements (Gupta et al. 2020). As with *Shielding* the ribbon is also impregnated with the SARS-CoV-2 RNA (coronavirus) from a plasmid construct.

Plague Votives

During 2020 I extended *Ex Voto* with a series of *Plague Votives* for an exhibition called "De Pest" (The Plague) at the Museum Het Valkof in Nijmegen, the Netherlands (Valkof Museum 2021). This show was several years in planning, a further indication that another pandemic was not completely unexpected, at least by the art community who are usually attuned to weak signals from the future.

Sublime Relics of the Laboratory

It is important to me to use an actual biological sample, even if in the case of dangerous pathogens those samples must be sterilized or made safe. The use of authentic materials creates a connection in the mind of the audience, inspiring curiosity and aesthetic reflection, and it does that for me too. I am always deeply aware of the biological materials I use and the awesome power they might have in their untamed forms. This integration of artistic and laboratory practice aims to shock viewers into an awareness of the social, ethical, and scientific issues from domestic violence to big data.

The incorporation of the invisible SARs-CoV-2 RNA from a plasmid construct in these artworks helps us to feel awe—a sense of the sublime power of the virus which has turned all our lives upside down.

The use of RNA in artworks made in this pandemic period is also a kind of genetic timestamp so that they become affecting secular relics of the sublime laboratory processes.

Notes

1 The artworks use SARS-CoV-2 RNA (coronavirus) from a plasmid construct. This is a safe, noninfectious reagent for SARS-CoV-2 research (NIBSC 19/304), obtained from the National Institute for Biological Standards and Control, UK. It is supplied to me by Ines Moura and Jane Freeman, who have been using this RNA for the development of a reverse transcription-polymerase chain reaction (RT-PCR) assay—a gene-based test—for detecting SARS-CoV-2 in feces.

2 *Shielding* was commissioned as part of the AHRC-funded ART/ DATA/ HEALTH: data as creative material for health and wellbeing, led by Aristea Fotopoulou at the University of Brighton (funded by the Arts and Humanities Research Council. AH/S004564/1 2019–2021). Anna Dumitriu's project is inspired by the work of community domestic abuse and violence charity RISE. info@artdatahealth.org http://www.artdatahealth.org @artdatahealth1

References

Defoe, D. ([1661?–1731] 2003). *A Journal of the Plague Year*. Penguin Books.

Dumitriu, A. (2012). "The Bacterial Sublime." *Anna Dumitriu*. https://annadumitriu.co.uk/portfolio/the-bacterial-sublime/

Dumitriu, A. (2018), "The Plague Dress." *Anna Dumitriu*. https://annadumitriu.co.uk/portfolio/plague-dress/

Gupta, H., et al. (2020). "Potential use of turmeric in COVID-19." *Clinical and Experimental Dermatology* 45(7): 902–3.

Ibsen, H. ([1828–1906] 1972). *A Doll's House*. Samuel French.

Oppenheim, M. (2020). "'Home became a prison': Calls to national domestic abuse helpline soar by more than third in pandemic." *The Independent*.

https://www.independent.co.uk/news/uk/home-news/domestic-abuse-rise-helpline-police-b1796937.html
Pepys, S. ([1633–1703] 1932). *Samuel Pepys' Diary*. De Luxe Editions.
Valkof Museum (2021). "De Pest." https://valkhofmuseum.nl/nl/bezoek/agenda/de-pest
Woolf, V. (1929). *A Room of One's Own*. Harcourt, Brace and Company.

20

Living in the Pandemic Panopticon

Who/What Is Watching You While You Think You Are Alone?

Karolina Żyniewicz

Coronavirus (Sars-CoV2), the leading player in the global pandemic declared in 2020, is everywhere, but it is impossible to say where exactly. I am writing that it is present because, even though the worldwide emergency state of the pandemic is over, the virus still exists and continues to kill (Tin 2023). My essay considers multiple time frames. Inspired by weekly virtual Viral Culture: Bioart and Society meetings initiated by Claire Nettleton in the spring of 2020, this essay was written the following spring through summer of 2021, and was revised with new perspectives in 2023. Exchanging observations with artists, scholars, and scientists from different corners of the world made me realize that despite cultural, political, and geographical differences, millions of people lived in some kind of Panopticon, where they felt observed and controlled: by the virus, other people, and by governments. The experiences differed depending on cultural and political backgrounds, but the common point was the same: control. As Danielle Couch, Priscilla Robinson, and Paul Komesarrof (2020) claim, the pandemic normalized surveillance systems in a Foucauldian manner. Before the vaccine became available, it was risky to touch and breathe, and COVID claimed millions of lives. Daily routines in many countries began to resemble one large laboratory full of biosafety protocols. Interweaving Viral Culture meeting discussions with theory by Michel Foucault, Jeremy Bentham, and Bruno Latour, and with my fieldwork as an interdisciplinary artist and scholar, and long-term project, *Signs of the Times*, I argue that the COVID-19 pandemic prompts new understandings of the construct of the Panopticon within the framework of a worldwide microbial outbreak.

Living in the Pandemic Panopticon: Who/What is Watching you while you think you are Alone?

It was the first wave of the coronavirus pandemic, which surprised most countries worldwide. Suddenly, a tiny agent, invisible to the naked eye (a virus), reminded us of our interdependence, physicality, and vulnerability. Governments faced the biggest global crisis since the Second World War. We observed chaotic struggles to maintain control by leaders, doctors, and finally, the virus. Our capitalistic freedom, or the illusion of freedom, became curtailed to save our lives through lockdowns and long-term confinements. After many years of almost unrestricted worldwide travel, we were asked to stay home. Distances became short—travel limited to the next neighborhood, or for the quarantined, the next room.

At the same time, the internet allowed us to remain connected with people far away. In March 2020, the virtual Viral Culture meetings brought together artists, humanities scholars, and biologists to reflect together on the situation. Besides exchanging information, we tried collectively to answer crucial questions, such as: How does the virus affect our understanding of humanity? How will the pandemic change our existence? How should we change our practices? By 2022, many aspects of life appeared to return to how they were before the pandemic, but these questions persist.

Every weekend, each participant opened an on-screen window, allowing everybody to enter one's private space and visit everybody else's private spaces simultaneously. We became flat and squared, without bodies, only virtual representations. Virtual communication platforms offered the option of viewing all participants in a tight grid or viewing the current speaker, a means to control whom and how we see. If someone participated without the camera, it was difficult to know if the person was truly present. People switching their cameras off went somehow out of the control of both moderators and other participants.

These same problems of "looking through the window" are related to all forms of online meetings, which initially occurred due to forced confinement, and have simply become daily routine. We were forced to be nosy, to look through windows, and see interiors, kids, and pets. Being inside and "looking in someone else's window" was a form of control, but a controlling performance also took place outside. Social distancing, wearing masks, lack of touch, and following these rules was not directly controlled by the police or government. People started to control each other. This became a matter of general responsibility (Suttie 2020). Humans could control each other, but they could not control the virus, which was supposedly everywhere. Before vaccines became available, Sars-CoV2 controlled human behaviors meaningfully. We felt "observed" by the virus at all times.

Biopolitical Panopticon

Observing and being observed recalls Foucault's idea of Panopticism, described in *Discipline and Punish: The Birth of the Prison* (1995), originally published in 1975, which begins with regulations set during the Black Death pandemic at the end of the seventeenth century, parallelling those implemented during the COVID-19 pandemic. The first standard control method in both cases was confinement.

> On the appointed day, everyone is ordered to stay indoors: it is forbidden leave to on pain of death. The syndic himself comes to lock the door of each house from the outside; he takes the key with him and hands it over to the intendant of the quarter; the intendant keeps it until the end of the quarantine. (Foucault 1995: 195–7).

However, although controlled isolation was crucial in both times, democracy brings a meaningful difference. Back then, control was, in some ways, more intrusive than in the case of the COVID-19 pandemic. The syndic, a government officer, was the one closing doors from the outside. He was responsible for keeping people isolated until the end of their quarantine. During COVID lockdowns, everybody closed the door by themselves, taking full responsibility for themselves and others.

> Only the intendants, syndics and guards will move about the streets and also, between the infected houses, from one corpse to another, the "crows," who can be left to die: these are "people of little substance who carry the sick, bury the dead, clean and do many vile abject offices." It is a segmented, immobile, frozen space. Each individual is fixed in his place. And, if he moves, he does so at the risk of his life, contagion, or punishment. Inspection functions ceaselessly. The gaze is alert everywhere. (195–7)

There is still law enforcement, but they react only in public spaces. In both cases, certain groups were afforded the privilege of risk, as government officials were allowed to be outside. However, it was a questionable privilege as they could be exposed to illness and death. Paradoxically, those closed in homes were more privileged, hidden from the virus and the controlling gaze, sometime evident, sometimes hidden.

> Every day, too, the syndic goes into the street for which he is responsible; stops before each house: gets all the inhabitants to appear at the windows (those who live overlooking the courtyard will be allocated a window looking onto the street at which no one but they may show themselves); he calls each of them by name; informs himself as to the state of each and every one of them—"in which respect the inhabitants will be compelled to

> speak the truth under pain of death"; if someone does not appear at the window, the syndic must ask why: In this way he will find out easily enough whether dead or sick are being concealed: Everyone locked up in his cage, everyone at his window, answering to his name and showing himself when asked it is the great review of the living and the dead. (195–7)

The action of window checking also took place in Poland (where I spent the first months of the pandemic), especially during COVID-19's first wave. The police were checking if people who were supposed to be in quarantine stayed home. The performance of appearing in the window worked somehow like virtual meetings on conference platforms. Every online gathering was started with questions: Can you see me? Do you hear me? If someone registered for a meeting and did not appear, the other participants could suppose that the person could be sick or have other pandemic-related problems. Appearing in the squared area on the screens of others, especially during online teaching, was a guarantee of genuine participation.

Doubtless, the most crucial practice, in case of both pandemics, was voyeuristic power. A common point is also that the danger is invisible, meaning that it is also uncontrollable: the virus (or in the case of Black Death, the bacterium). Humans are unable to control the virus, so they try to control everything else. The virus becomes visible only as symptoms of infection or rather disease because some people can be infected without any symptoms. We can see the virus only through technological extensions, such as microscopes. Seeing the virus is possible only in laboratory conditions, where microscopes render it visible. However, even in the laboratory the virus can easily go beyond our control. In biological laboratories, control is based on following biosafety protocols. That is why it seems obvious that in order to try to control the virus outside the lab, we need to apply biosafety protocols outside the laboratory too.

An essential part of laboratory control is sterility, related strongly to the idea of cleanness. In contemporary laboratories, objects and surfaces are disinfected with ethanol or by UV lightning ("Biosafety" 2022). This alcohol disinfection had to be extended to the reality outside the laboratory. In the seventeenth century, the idea of sterility did not exist or at least was understood differently.

> Five or six days after the beginning of the quarantine, the process of purifying the houses one by one is begun. All the inhabitants are made to leave; in each room "the furniture and goods" are raised from the ground or suspended from the air; perfume is poured around the room; after carefully sealing the windows, doors and even the keyholes with wax, the perfume is set alight. (Foucault 1995: 195–7)

The purifying procedure described shows an interesting battle with the plague's invisible actor. The pathogen is fought with perfume. Supposedly, perfume was then produced based on alcohol so the intuition was correct. Additionally, a pleasant smell could reduce stress and fear and give a feeling of being safe again.

In "Give me a Laboratory and I will Raise the World" (1983), Latour wrote about the process of laboratorization, giving Louis Pasteur's anthrax research as an example. Pasteur had to find a way of applying his research to the reality outside the lab, which he did by creating the vaccine. Anthrax became an important social agent having influence on agriculture. An analogous process takes place with coronavirus: it became a social and cultural rule-setter. As scientific investigations about the virus are happening inside and outside of the lab, reality is a crucial area of experimentation. Activities before associated only with laboratories and hospitals have been transferred to our daily routine, especially wearing masks and gloves and using sanitizers. We have all become lab rats of sorts, without our consent, but for our own good.

Foucault references Bentham's concept of the Panopticon, an architectural project of centralized control. This project was intended mostly as a prison but also possibly a psychiatric hospital. Instead of describing construction details, I will focus on the main idea: "Visuality is a trap" (Foucault 1995: 200). In the Panopticon's case, the prisoner/patient is controlled by being visible: "He is seen, but he does not see; he is the object of information, never a subject in communication" (200). The person being seen is objectified. During the Black Death, the situation was similar. People were confined to

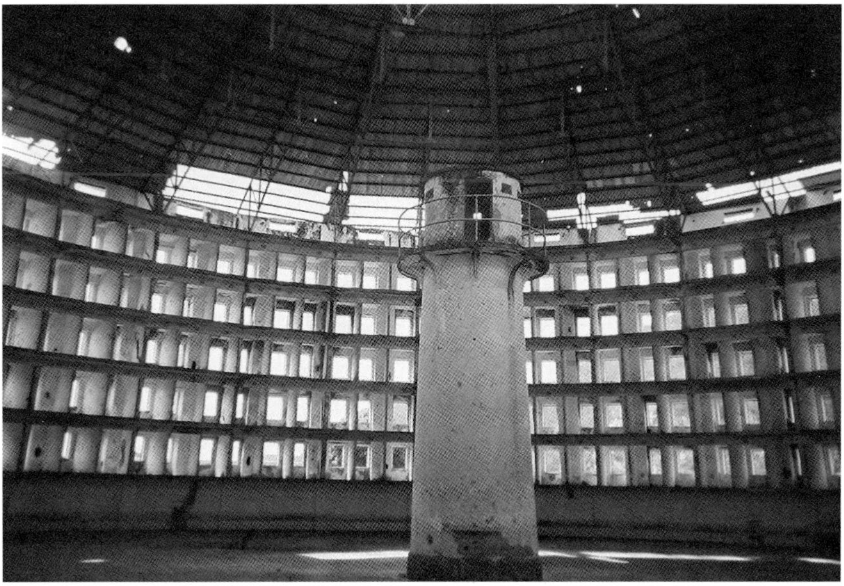

FIGURE 20.1 Inside one of the prison buildings at Presidio Modelo, Isla de la Juventud, Cuba, *2005. Courtesy of Creative Commons.*

their homes by guards. Choosing to follow the rules was not an individual decision, as syndics kept their house keys. The coronavirus pandemic has a different background, since in contemporary society everybody is responsible for respecting rules. Disobedience can be penalized, but it is a matter of individual decision. In Bentham's 1791 project, prisoners/patients could not be sure when guards were actually observing them; thus they felt they were always being watched. They needed to behave properly, just in case. Can we say that the virus functions as a kind of a guard? Can the virus see us?

Nonhuman Guard

In biological nomenclature, a virus is not even an organism. Viruses are submicroscopic particles that can infect living cells. They are unable to control their direction of movement, but are able to control us without this function. Similarly to the Panopticon's guards, we do not have to see them to suspect they are there—that is, everywhere. The invisibility of the virus has a greater controlling power than the visibility of police because a financial penalty is nothing compared with illness or death. Knowing the virus is everywhere, all we can do is try to control its transmission. The tricky thing is that the only direct evidence that someone is the carrier is the disease itself. To detect nonsymptomatic carriers, we need to use medical testing. The results of this testing do not make the virus visible. We receive a piece of paper or rapid test saying we are positive or negative, wherein being positive means one is a carrier of the virus. The virus remains hidden.

Signs of the Times: Collecting Biological Traces and Memories

To add a fieldwork element to this text I will describe my pandemic-linked project under the working title, *Signs of the Times. Collecting Biological Traces and Memories* (Zyniewicz 2021), related to the biopolitical Panopticon which we have all lived in since March 2020. It touches upon the virus's invisible omnipresence and the irony of biosafety protocols. Masks, a key feature, became an element of our daily outfits. Disposable masks were to be worn only once. However, it was hard to find anyone strictly following this. Usually, people used the same mask at least twice. Microflora coming from our breath settled inside the mask and we inhaled it again. Streets and parks became full of discarded synthetic medical masks. I saw an announcement that we should throw masks into rubbish bins only having placed them inside a plastic bag beforehand, supposedly to protect cleaning staff. The virus can survive on inanimate surfaces for days (Nature Reviews Microbiology 2021: 148). People cleaning the streets wear gloves. It made me think about this panoptical aspect of the pandemic. The announcement's writer was convinced that masks on the ground are potential sources of infection. You never know where the virus is and how long the mask has been lying there.

The announcement inspired me to play with rules and social imagination about the virus. I decided to collect masks found in different places in Berlin. This collecting had an experimental and performative character. I was testing my own fears and presumptions as well as people's reactions seeing me pick up dirty masks from the ground. Every mask I found got me thinking about the person who was wearing it. Were they a carrier? Maybe it was owned by someone sick? What if they had some illness even more serious than COVID-19? A bigger problem than these uncertainties was being ashamed. I was ashamed to be seen by other people in the park. I was afraid someone would ask me what I was doing, or think I should be locked up. You could say I felt seen even when nobody was looking at me. I was more afraid of people's reactions than the possible presence of the virus on the mask.

The second stage of my project will be based on laboratory examination. I am imprinting the masks on a medium's surface, allowing whatever organisms exist on the masks' surfaces to grow. The nonhuman pandemic participants will be identified. The first laboratory experiments have been done already. Fortunately, I have not gotten infected or fallen ill. Does it mean I have not encountered the virus? Did I touch it while touching the masks? Maybe it was not there, where I expected it to be.

Visibility is not only the tool of control but also a form of evidence, which helps us to believe, or rather to be convinced. The virus is invisible without the microscope, so there are those who deny its existence. This is an interesting group in the context of the biopolitical Panopticon. They try to flee from under the control machinery, refusing to follow biosafety protocols, including wearing masks. It does not mean they are not seen/observed, but they pretend they are not or do not care. In context of my project, it is hard to say if the reaction of pandemic denialists to my mask collecting was different from the reaction of those convinced that the virus lived on the mask. Technically, nobody reacted openly to my actions; I did not notice any strange looks. The "guard" was only in my imagination.

In a further stage of the project, when gathering started to be possible again, I decided that anonymous masks were not enough. I wanted to see the mask carriers and hear their stories. In the first version of mask and memory collecting events, I asked people to sit in front of the camera, and in front of me (the camera located between us). I gave them a fresh medical mask for when they told me their pandemic memories. Gazing was evident in this situation. I was sure that after so many months of remote meetings with a camera on, talking to a physical camera would not be a problem. I was wrong. The camera was a block for some people, who felt like participants of a TV show. My presence behind the camera helped, because then they

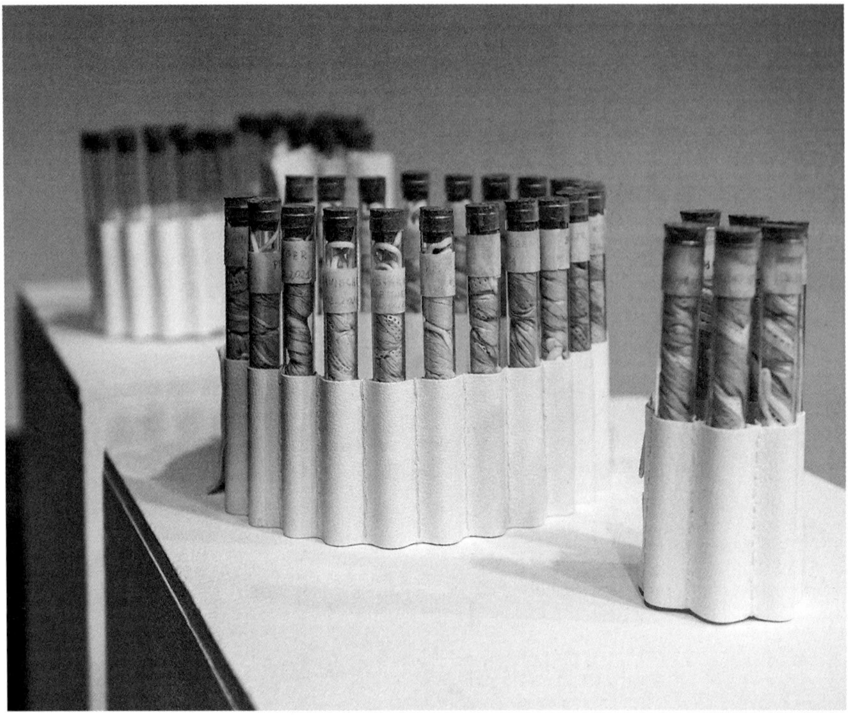

FIGURE 20.2 *Karolina Żyniewicz,* Signs of the Times. Collecting Biological Traces and Memories, *documentation of the event at Contemporary Art Centre Łaźnia in Gdańsk. Photo by Adam Bogdan. Courtesy of the artist.*

could look into my eyes, but the technological eye was still a burden. To use the recordings in the future, I had to ask participants to agree to be seen by gallery visitors. People who wished to share private and complicated experiences felt especially uneasy being visible. Covering faces with masks did not help much. Considering the problems with visibility, I decided to eliminate vision from my project and focus on sound recordings and transcriptions. Visibility will be reserved for the nonhuman beings growing from the mask samples. It will somehow subvert the usual order.

Staying in Touch by Seeing

In the concept of the Panopticon, prisoners and patients cannot see their neighbors or communicate with them. It is lonely even under constant surveillance. Nothing redirects attention away from the idea: the guard sees me, I need to behave properly. The lack of interaction between inmates is a tool of control (Foucault 1995). By talking to each other and building relations, they could have motivated each other toward disobedience. Interacting makes the situation of captivation easier to bear but harder to control. Isolation is, in this concept, a basic element of punishment. In the biopolitical pandemic Panopticon, the situation was a bit different. Of course, lockdown and quarantine meant physical isolation but the internet allowed us to stay in touch by seeing each other. In this context, seeing and being seen is not the tool of repression, but rather something that helps us survive. As I stated before, this is looking through the window, but also a chance to talk and to support each other in crisis.

The Viral Culture Bioart and Society meeting group, which I mentioned as an inspiration, was a coalition of window prisoners helping each other survive physical separation. Every participant could share what was happening in his/her/their "cell." The discourse was a meta-narration on the pandemic's biopolitical Panopticon. Through this volume, it is also a unique record of how the global art and science community built a narrative around the pandemic, trying to understand the situation, but also finding ways to imagine what we can learn from it.

References

"Biosafety: Decontamination Methods for Laboratory Use" (2022). https://blink.uscd.edu/safety/research-lab/biosafety/decontamination/index.html

Couch D. L., Robinson, P., and Komesaroff, P. A. (2020). "COVID-19: Extending Surveillance and the Panopticon." *Journal of Bioethical Inquiry* 17(4): 809–14.

Foucault, M. (1995). *Discipline and Punish. The Birth of the Prison*. Vintage.

Hu, B., Guo, H., Zhou, P., and Shi, Z. L. (2021). "Characteristics of SARS-CoV-2 and COVID-19." *Nature Reviews/Microbiology* 19: 141–54.

Latour, B. (1983). "Give me a laboratory and I will move the world." In Knorr-Cetina, K., and Mulkay, M. (eds.), *ScienceObserved*. Sage, 141–70.

LibreTexts: Biology. *Viruses characteristics*. https://bio.libretexts.org/Bookshelves/Introductory_and_General_Biology/Book%3A_Introductory_Biology_(CK-12)/07%3A_Prokaryotes_and_Viruses/7.07%3A_Virus_Characteristics

Suttie, J. (2020). "Five Ways to Encourage Safe Behavior During the Pandemic." *Greater Good*. https://greatergood.berkeley.edu/article/item/five_ways_to_encourage_safe_behavior_during_the_pandemic

Tin, A. (2023). "COVID Global Health Emergency is Officially Ending, WHO Says, but Warns Virus Remains a Risk." *CBS.com*.

Żyniewicz, K. (2021). *Signs of the Times. Collecting Biological Traces and Memories*. Video. https://www.youtube.com/watch?v=YGcTxSnpW38

21

Viruses as Testing Grounds for Speculations

Pei-Ying Lin

Interactions between humans and viruses happen not only at the level of biological physicality, but also at the level of human consciousness—we are infected by the virus through both physical and perceptual encounters. These interactions induce an angst toward uncertainty when confronted with the notion that viruses inhabit our bodies. This phenomenon becomes extremely fruitful when implemented in an artistic context. This chapter explores this phenomenon through two participatory artworks: the cookbook, video, and performance work, *Virophilia*, and the story workshop, *Discourse of a Viral Boundary.*

In *Discourse of a Viral Boundary*, fictional viruses become tools for "thought experiments." Through character creation and storytelling, participants explore atypical scenarios in human–virus interaction—as opposed to the public health approach of seeking general principles. Role-playing a "pandemic-storytelling" of fictional viruses becomes a platform that empowers participants to reflect on the pandemic themselves at different levels. In *Virophilia*, the concept of the virus is considered through its "appreciated features" rather than its "undesirable existence." Fictional recipes make use of viral features as different culinary methods, which are further explored in a dinner performance to create embodied viral experiences.

In the time of COVID-19, these artistic approaches engage people in alternative conversations around viruses that opens their minds to the practicalities of viruses and the possibilities of human–virus relationships.

As human society navigates the waves of the COVID-19 pandemic, we are suddenly reminded of the long-forgotten coexistence between humans and viruses. This opens up an interesting landscape of human consciousness that is often ignored by modern metropolitan life. Human society has never been free from pandemics; we have constant interactions with viral others, some of which are even asymptomatic. During COVID-19, the fear of a collapsing medical system urged nations to implement different methods that requested their citizens to modify behavior, such as social distancing and quarantine, unfolding sophisticated individual narratives. Due to the invisibility of viruses, and the latency of virus detections—where even rapid tests take 15 minutes to reveal the virus's existence—we engage in constant speculation about our encounters with coronavirus. The interaction happens not only at the level of biological physicality, but also at the level of human consciousness—we are infected by the virus through both physical and perceptual encounters, where the virus pervades our imagination.

Viruses as tiny, omnipresent, penetrative, biological entities push into our minds and we suddenly see things in new ways. Although nature–culture dualism has dominated Western thought, as if human culture is independent from the happenings of nature (Haila 2000: 155–75), the COVID-19 pandemic had proven such separation is problematic. This urges us to humbly explore the meaning of viruses to humankind.

In the past ten years, my artistic practice has explored different topics related to viruses, through a fascination that their existence overrides our social norms and interacts directly with humanity at a biological level that is yet too sophisticated for us to comprehend. Although there are studies and statistics, our individual reaction toward a specific virus will only be revealed when we become infected, which creates an angst toward the uncertainty we face when confronted with the possibility of viral infections.

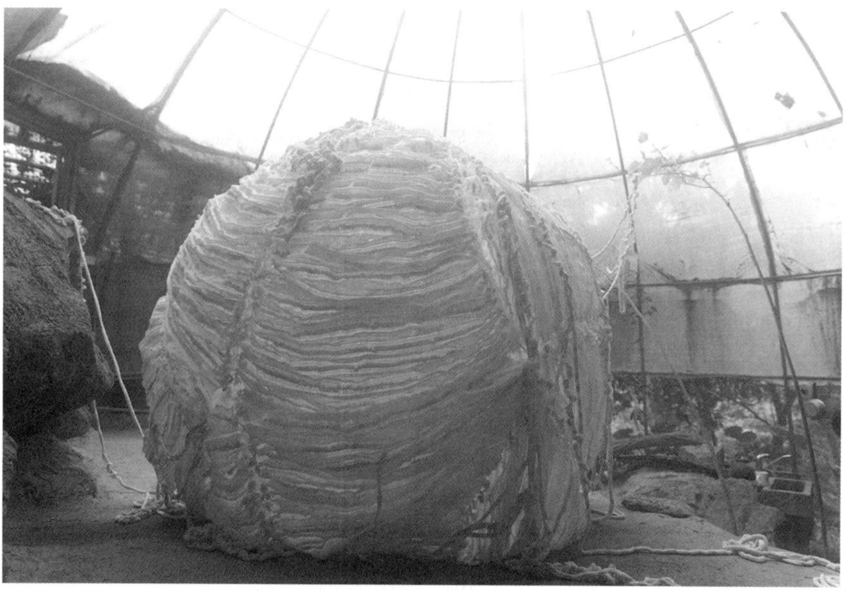

FIGURE 21.1 *Pei-Ying Lin.* Studies of "Interbeing—Trance 1:1," in the process of becoming part of "the Nature." *Courtesy of the artist.*

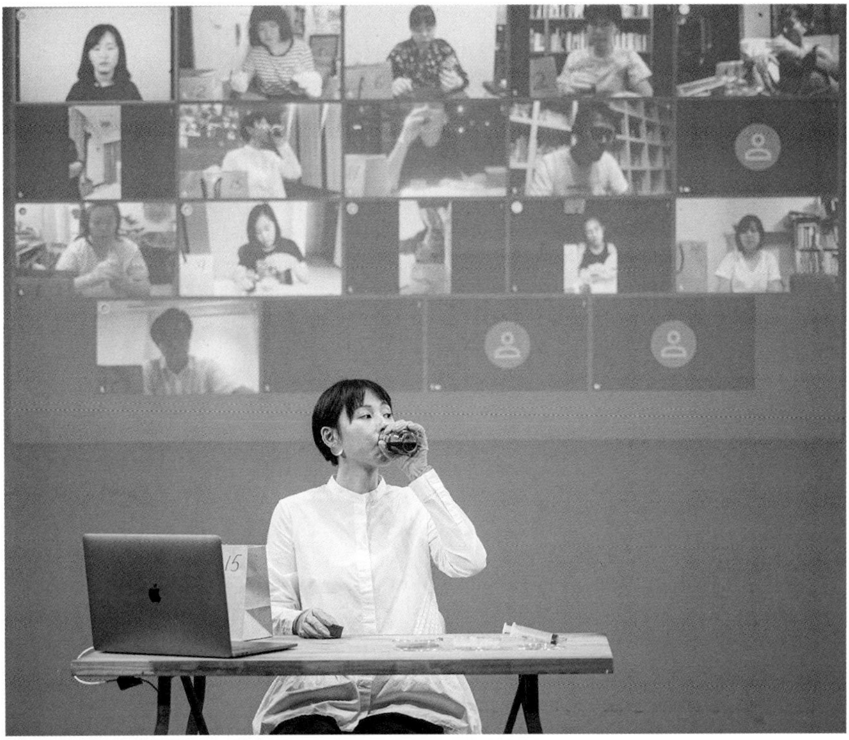

FIGURE 21.2 *Pei-Ying Lin,* Virophilia, *Dinner Performance, 2020 Photo by C-Lab Taipei. Courtesy of the artist.*

This phenomenon becomes extremely fruitful within art practice. I will discuss my projects *Virophilia* (a speculative cookbook, video installation, and series of food performances) and *Discourse of a Viral Boundary* (collaborative storytelling workshops) as examples.

In *Discourse of a Viral Boundary*, which was developed during the first year of the COVID-19 pandemic and conducted between 2020 and 2021, participants are invited to have a two-with-one session, where I serve as a guide and provide them with a narration tool to construct possible infectious disease outbreaks (Figure 21.1). The two participants are first asked to choose four characters from the characters' deck—which are drawn as semi-human forms—and create the characters' background, personality, physicality, and mentality. One of the characters is given symptoms surrounding a body part chosen by the participants, and a transmission route that was assigned randomly. The pandemic narration begins. Participants are asked to speculate how these elements of viral characteristics—such as fever or sore throat—will interact with our society, through four stages of escalation, with society eventually reaching the point of no return. New interaction restraints are in place. The physical conditions of the majority are changed. Political powers are shifted. And the participants are asked: "How do the characters adapt to this new norm?" For example, one group of participants speculate a fictional sexually transmitted disease that heightens sexual desire, as it spreads the majority develop a heightened sex drive, which makes public opinion orient toward sexual openness. The tool motivates participants to think about the opportunities that chance circumstances bring and engages them by alternating the narrative perspective between the human and the virus. After each stage of escalation, participants are asked to respond to the spreading strategy from the virus's point of view, which launches the story to the next, more intense stage.

The components of the narration tool were developed based on the pattern of a pandemic, listing crucial elements that are often examined in public health sectors. The tool consists of a form that is divided into stages, with cards for characters, symptoms, and transmission routes. However, what differs from epidemic investigations is that the "pandemic" in this case was fictional and unfolds according to the participant's will. Instead of passively experiencing how a pandemic unfolds or being in the passive position of examining historical accounts, the participants are given an active role to create the story, broadly following biological principles.

Very often, participants put a bit of themselves or someone they know into the characters and the stories. This semi-attachment frees them to talk about their feelings and thoughts during the pandemic from the character's perspective, allowing them to experiment with the "what if" of seemly incorrect decisions, eliciting responses such as, "perhaps the character will enjoy lockdown because that's when she finally has time for herself"; "it would be really nice for the characters to give more care to each other, that will help with what's happening"; or "I think it is quite normal that people will have discriminations, and these discriminations can be very misleading sometimes. Let's play with one of the elements in this story."

In some cases, participants' personal experience echoed the scenarios they described. Some of the characters created by participants were put to death in different ways by the pandemic as a reflection of what happened in the real world. For instance, among the characters participants created, there was one with schizophrenia, who died during lockdown as his caretakers were not able to check on him. He committed suicide because he forgot his medicine. Another character, a lady who was infected with a virus that causes kidney failure, decided to use her sickness as her weapon and infected her ex-husband who abused her. A third character, a transgender priest, moved to Taiwan from Austria because they could not find answers to their questions in the Bible anymore. When the fictional virus infected them, they found their newfound immense sexual desire—which was beneficial for the virus to spread—difficult for their position as a priest. In the same story, another character, who is a social researcher, found the sex-eager symptom challenged his definition of morals. The story ended with the society divided into two parts: the conservatives, who are not infected because they are sexually conservative, and the liberals, who are infected and seek sexual liberation.

Although conceived earlier, these narrative sessions were conducted during the COVID-19 pandemic, and with the first version of the narrative tool, many pandemics started with coronavirus symptoms as if the only way we see sickness is through fever and respiratory difficulties. With the second version of the narrative tool, participants were asked to choose a body part before coming up with the symptoms, which lead to a greater diversity of sicknesses. That was when the stories began to show a good mix between having a biological "skeleton" according to the principles of a virus and having the social-cultural "muscles" of human behavior, leading to the aforementioned interesting comments from participants. These "thought experiments" with fictional viruses that participants engaged with explored atypical scenarios in human–virus interaction as opposed to the public health approach of seeking general principles.

The narrative tool, along with excerpts of stories, was presented at a 1.5-hour workshop at Zone2Source, Amsterdam, as a part of the launch event of the School for MultiSpecies Knowledges (Shapiro 2021). The methodology was discussed and the notion of "symptoms" was deeply considered. The original narrative tool started with a semi-linear structure that referenced the physical world. However, during the workshop participants mentioned that despite many not being infected by the SARS-CoV-2 virus directly, there was a sense that we all had the "pandemic symptom" from day one when lockdown started. The psychological aspect of our wellbeing was deeply influenced by the conceptual existence of SARS-CoV-2, and we were fundamentally changed by the presence of it. With regards to the physical body, there was the layer of typical biological symptoms that we are familiar with, but also the layer of nonsymptoms—for instance, sore throat and fever, that were not apparently caused by the virus, yet which we suspected were due to the virus when tests were unavailable. There is also a layer of asymptomatic symptoms, where one believes they are not infected because there are no symptoms, but still sheds the virus. At the conscious level, we are also engaged with symptomatic, nonsymptomatic, and asymptomatic expressions of our conscious behaviors as a "human collective." At the individual level, having emotional and psychological reactions toward the pandemic is the symptom, while similar reactions toward things that are not directly linked—yet thought to be linked—to the pandemic is a nonsymptomatic characteristic, and transmitting information and thoughts about the pandemic without being influenced oneself, is asymptomatic.

To further investigate the notion of symptoms by looking at all the strains of SARS-CoV-2 as one collective entity, and human society as the other, the "symptoms" at the scale of human society are disruptions in international flight travel, nationwide lockdowns, excessive cargo transportations, and a shifted online presence. This extends further if you are unlucky enough to be in those areas where politics exploit the pandemic—for instance, Slovenia, whose right-wing government took advantage of the pandemic lockdowns, or Taiwan, whose position on pandemic issues became an international political wrestling between China, Japan, and the United States. These societal symptoms of COVID-19 are extremely dynamic, even though, in a physical sense, the whole world appears to be sitting still.

Discourse of a Viral Boundary revealed a spectrum that is only obvious when the virus becomes a tool for discussion and reflection. Because viruses are biological, and have coexisted with us as long as humans have existed, our reactions toward viruses are almost instinctive, although sometimes problematic, but never "un-noticed."

The characteristics of viruses are explored in a neutral way in the project *Virophilia* with their functionalities as "appreciated features" rather than an "undesirable existence." *Virophilia* is a cookbook written in the year 2068 by a fictional author who only reveals her initials as P.-Y. L. Retrospectively, the book explores the development of including viruses in culinary cultures as beneficial—as a fetish, a boost to immunity, or a form of vaccine. The "trend" begins with simulating the sensation of sickness in dishes without using actual viruses, then moves on to using viruses for fermentation—as we do today with yeast and bacteria—and then to using biotechnologically modified viruses/vaccines that trigger short-term sickness in up to a few hours as an added "spice" of the dish. The book ends with dishes that are designed for the whole food chain with viruses as the triggering point.

The specific characteristic of a virus that lends itself to this speculative artwork is invisibility. Apart from the cookbook, the project was also presented as dinner performances at different locations (Figure 21.2 and 21.3). Selected

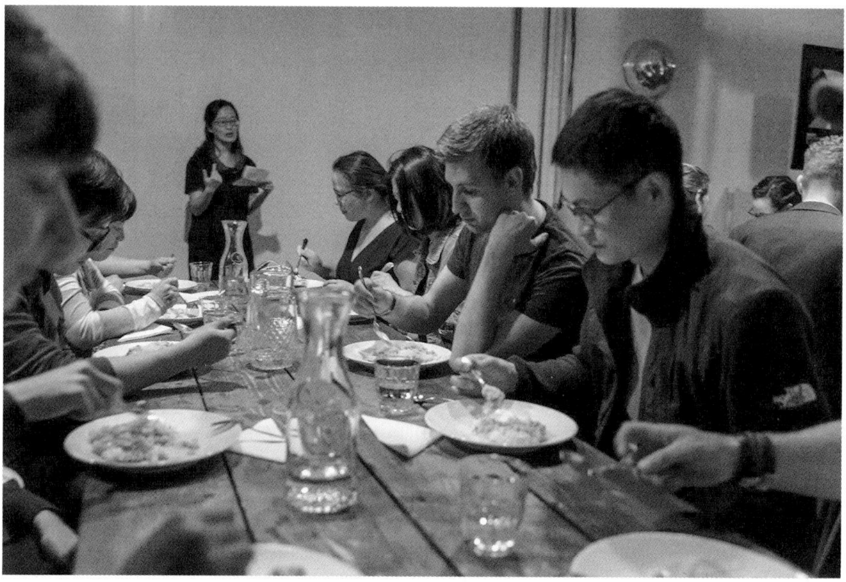

FIGURE 21.3 *Pei-Ying Lin,* Virophilia, *Dinner Performance during Cinemasia, 2019. Photo: Armando Ello. Courtesy of the artist.*

dishes, and newly invented dishes following the core idea of the cookbook, were provided at the performance, where the participants ate while listening to stories that detail how the viruses from the dishes enter and interact with their bodies. As a speculative work, viruses were not deliberately incorporated into the recipes, but physical sensations were emphasized using ingredients that trigger similar body reactions, causing a semi-placebo effect. Because viruses are invisible and do not have an immediate effect on their hosts, the latency of physical responses that we have learned since we were born presents a space between "viral encounters" and "conscious acknowledgments of an infected body" that offers enormous flexibility to play with. If corresponding to the layers in *Discourse of a Viral Boundary*, *Virophilia* offers a space for nonsymptomatic phenomenon. The difference between the two projects is that in *Discourse of a Viral Boundary*, the human–virus relationship could be either beneficial or harmful, pleasant or unpleasant; but in *Virophilia*, the harmful is reframed as beneficial, and the unpleasant is put into a context where it is perceived as pleasant. The context of art also allows both the creators and the participants to extend beyond existing parameters of what is possible, while being aware that these manipulations were only valid within the projects. Once the participants "exited" the project—when the performance ends or when the workshop ends—they were immediately able to compare the fictional narration and the science-bound reality which allows participants to be critical about the current state we find ourselves in and experience the importance of interpretations and perspectives.

The fundamental elements that inspired the fictional narrations in *Virophilia* and *Discourse of a Viral Boundary* were mainly built upon past human experiences with viruses. No one in this world has been exempt from viral infections, and these infections generate a shared memory of how sickness feels, how abstract infection can be, and how communicable diseases spread. These experiences are multisensorial, common in all cultures, and apply to all races. These shared memories are also reminders that we have always been engaged with viruses no matter how much we ignore this fact. At the same time, virus–society interactions are extremely sophisticated and we tend to accept that narratives in these contexts can be "unpredictable." The combined result of shared memories and acceptance of unpredictableness, together with a given artistic context, gives us a neutral standpoint where we gain the ability to reflect upon ourselves as human, as if viruses are our mirrors.

References

Haila, Y. (2000). "Beyond the Nature-Culture Dualism." *Biology & Philosophy* 15(2): 155–75. https://doi.org/10.1023/a:1006625830102

Shapiro, I. (2021). "School for MultiSpecies Knowledges, Zone2Source, Amsterdam, The Netherlands." https://irinashapiro.nl/school-for-multi-species-knowledges.html

INDEX

Entries followed by *f* indicate a page with a figure.